S0-BWX-239

DESIGN FOR COMMUNALITY AND PRIVACY

DESIGN FOR COMMUNALITY AND PRIVACY

Edited by Aristide H. Esser

The Association for the Study of Man–Environment Relations, Inc.
Orangeburg, New York

and

Barrie B. Greenbie

University of Massachusetts
Amherst, Massachusetts

PLENUM PRESS · NEW YORK AND LONDON

Library of Congress Cataloging in Publication Data

Environmental Design Research Association.
Design for communality and privacy.

Includes bibliographical references and indexes.
1. Dwellings—Congresses. 2. Environmental engineering (Buildings)—Congresses. 3. Personal space—Congresses. 4. Crowding stress—Congresses. I. Esser, Aristide H., 1930- II. Greenbie, Barrie B. III. Title.
TH4812.E57 1978 301.5'4 78-7055
ISBN 0-306-40010-3

Based on a workshop convened by Aristide H. Esser at
the Sixth Annual Conference of the Environmental Design
Research Association held in Lawrence, Kansas, April 21, 1975

A Division of Plenum Publishing Corporation
227 West 17th Street, New York, N.Y. 10011

Printed in the United States of America

Preface

This book developed from the workshop, "Design for Communality and Privacy," convened by Aristide H. Esser at the 6th Annual Conference of the Environmental Design Research Association in Lawrence, Kansas, April 1975. Since the late sixties, groups of behavior scientists and designers have been trying to engage in a common effort to improve our built environment. Thus, when in the 1974 meeting of the American Psychological Association such concepts as territoriality, privacy, personal space and crowding were discussed,[1] the logical next step appeared to be translation of behavioral findings into design recommendations. Most of the EDRA-6 workshop papers addressed these issues from a viewpoint potentially useful to designers, and these are included in the book. Subsequently, however, some papers were specially written for this collection to provide more balance in the range of topics. When a publisher was found, Barrie B. Greenbie joined as Editor to cover design issues.

We thank all contributors for their efforts which made this book possible. Special thanks go to Alton J. DeLong and his secretaries, Debbie Goldsmith and Tina Breegle, as well as Loretta Kaufman, who typed the manuscript. We are grateful for the assistance and patience of Seymour Weingarten, Editor, and Stephen Dyer, Senior Production Editor, at Plenum Press.

Aristide H. Esser
Barrie B. Greenbie

[1]Gary W. Evans and Daniel L. Stokols (Eds.), Special Issue on Privacy, Territoriality, Personal Space and Crowding, *Environment and Behavior*, Vol. 8, No. 1, March 1976.

Contents

INTRODUCTION

Aristide H. Esser

Association for the Study of Man-Environment Relations
P. O. Box 57, Orangeburg, New York 10962

Barrie B. Greenbie

Department of Landscape Architecture and Regional
Planning, University of Massachusetts
Amherst, Massachusetts

The papers in this volume reflect the increasing joint efforts of behavior scientists and designers of the built environment to organize space in a socially meaningful way without stunting conceptual growth. The concepts of communality and privacy in most of the following papers originated as verbal presentations to provide a renewed basis for such efforts. In a conference with face-to-face exchanges, it is possible to discuss such concepts meaningfully for a particular built environment; most designers implicitly recognize that the sense of one's self as a distinct entity and one's sense of belonging to a social group are universal. However, because concepts of self and community are so central to the experience of living, they are extremely difficult to pin down formally.

Any attempt to deal with them explicitly, as the conference which generated these papers attempted to do, must face the often tedious and always imperfect process of defining terms. Spoken language is so flexible and can be attached to so many different realities precisely because that portion of the human brain governing speech is a latecomer in evolution. Scientists, like lawyers and others who organize experience in terms of verbal thought, must redefine their terms with each new insight they have into an infinitely complex reality. But language, the primary tool for human analytical thought, cannot deal satisfactorily with many realities of human experience that reach back into preverbal spatial experience.

Now, physical designers do not always share the need to define terms as do scientists. They tend to think visually and spatially and use words in somewhat more loose terms, specifically for their power to evoke a spatial image. Since we hope that the following pages will ultimately be assimilated by designers, we ask their indulgence in following thought systems that may appear irritatingly opaque. Each scientist trying to address a multidisciplinary audience has, of course, clear definitions in mind (whether or not they are stated explicitly) which may be shared by those in his area of knowledge, but may not be at all clear to others. In introducing the authors of these pages to a design audience, we shall attempt here as best we can to provide a general definition of the subject terms of this book.

Communality and privacy are not separate concepts, but corollary ones, essentially expressing different aspects of a single experience or feeling. Generally speaking, privacy pertains to the sense of self or of intimacy with others, whereas communality pertains to the sense of being at home in a group, or among many others. But one cannot have a sense of self without an equal and complementary sense of others. Individuality and social belonging thus are both expressions of social behavior. The problem of recognizing this inherent dualism is complicated in that most modern human beings do not live in a single social context or group, but move in and out of many different social settings.

Because all behavior takes place in some sort of space, social identity and social behavior cannot be separated from physical space. Some of the theoretical background and the implications of the use of space in one way or another as part of behavior have been described in a previous conference on the Use of Space by Animals and Man (5). The use of space as a measure of the behavior of animals is central to the young discipline of ethology, which has brought to behavioral scientists and designers alike such important concepts as "territory." Territory is a means of assuring the animal survival and identity. This includes, among social animals, the use of small group communal space, particularly by a breeding unit. We assume that in human beings territory likewise is an expression of these oldest biological impulses. For the individual, territory may be related to privacy; for the primary group, it may relate to community feelings. These ideas are more fully developed in the first paper of this volume by Esser.

Those physical spaces which facilitate the impersonal secondary group relationships, so characteristic of commerce and technology, may be less consistently considered in terms of territory. Such relationships may, to varying degrees, reflect a larger level of communality and often are not clearly attached to specific spaces. This may be attributed to the human ability to organize social behavior conceptually by means of prosthetic symbols, using verbal and mathematical language. However, even primary group territories serving feelings of community often need to be bounded and defined by abstract symbols such as signs saying "No Trespassing" or "Club Room." For designers, it is important to recognize the intense meaningfulness of seemingly absurd physical symbols, such as a small picket fence, which can easily be stepped over but which nevertheless may constitute a strong social boundary.

The problem of clearly defining the concepts of privacy and communality is complex because in a very real biological sense we live on different levels of experience, both socially and spatially, due to the way evolution has organized our brains. (Several chapters in this volume elaborate on this issue.) Our scientific knowledge indicates that human social-spatial behavior emerged in small tribal groups in which members know each other intimately. In such groups there is less explicit need for concepts of community, communality and privacy, since personal identity is so closely related to tribal identity and this remains true of many sub-cultures today. In such primary groups, some spatial privacy is provided for some personal functions, but personal identity, cultural identity and spatial identity tend to be the same. This does not preclude the fact that early human societies were highly mobile; the concept of territory does not always posit a fixed geographical space. The markings evolved by different animal species for territories are considered to be a consequence of the fact that animals move, and mankind in particular has littered the world with territorial symbols.

When human beings learned to domesticate plants and animals, they actually became more, rather than less, dependent on maintaining social territories fixed in space. On the other hand, as surplus food made possible the development of differing roles for individuals within a group, the identity of the *individual* as a person able to move from one group to another and to enter into "social contracts" with individuals from other groups becomes increasingly significant. At the same time, more impersonal and ephemeral secondary group relationships become increasingly important to human survival. Some individuals retain their emotional attachment to the small, essentially tribal, primary group, while others transfer emotional allegiance to roles, organized in conceptual space rather than ethnic groups defined by place. For the latter, spatial arrangements that facilitate secondary group activities and free movement across boundaries become extremely important. For these, privacy tends to become an individual matter, or at least the province of a group no larger than a nuclear family, while communality is associated with abstract symbols, often shared by people who are emotional strangers.

Nevertheless, for large portions of the world's populations, the familiar primary group of the extended family, or a surrogate family of friends and neighbors, remains crucial for both identity and security. In distinguishing between these qualitatively different kinds of privacy and communality, which will be elaborated further in the last chapter of this volume by Greenbie, it is important to consider the concept of "context," as advanced by Hall (8), which will be explored more fully in a number of the following chapters.

The complexities of designing for communality as a context for community and privacy in democratic and pluralistic societies runs into sharp conflict with legal structures based on an oversimplified model of these relationships. Especially in the United States, laws designed to serve equality of access to space, resources and freedom of life style come into collision with rights to ownership of land. Generally speaking, the law in the United States and similar countries makes no distinction between privately owned property used for primary group purposes, such as residence, and that used for essentially secondary group purposes, such as profitable commerce. It permits no distinctions between rights of access to a small neighborhood park which may serve as a community focal point for a small ethnic enclave and a large public park serving an entire city. The designer, even if he is successful in understanding and translating to spatial terms the findings of behavioral science, may be prevented from implementing his understanding by relatively crude and arbitrary zoning laws, land-use plans and powers of eminent domain (7, Chapters 10 and 11). The well known financial constraints which face designers are, of course, functions of the legal structures that control economic behavior.

Two other aspects of this complex subject, which will be explored in various ways in the pages that follow, are tnat the design of space, from a behavioral point of view, must accommodate action chains, which, as noted, involved first-hand communication links between individuals playing different roles. Even on the level of such abstract secondary groups as multi-national corporations, face-to-face contact remains essential. On the other hand, if common spaces are designed for communality between people from differing groups, the design must provide opportunity for what Coss calls "cut-off acts" (4). These enable a person not to have eye contact with another and still show communality. An example might be wall hangings in an interview room which give the person being interviewed an excuse to look away without appearing to be unfriendly or evasive. The other related aspect is that action-chains, whether personal or impersonal, require designing space to permit different functions and express different characteristics of social behavior over time (9, 10).

Efforts to define and otherwise come to terms with the

pervasive but elusive concepts of communality and privacy have generated an increasingly rich body of literature. Much of the initial interest came from architects and planners (1, 3, 6). But recently a large portion of books and articles have come from anthropologists, sociologists, psychiatrists, psychologists, etc. Some recent books deserve special mention here.

One of the most influential writers is the anthropologist, Edward T. Hall, who coined the term, "proxemics," to describe the culturally determined (primary group) use of space as a communication system. His new book, *Beyond Culture*, (8) pays particular attention to cultures which experience the environment holistically. He calls these "high-context" cultures as compared with those dependent on high information situations, typical of many technological environments, where unrelated bits and pieces of information tend to overload the information processing capacities on the Central Nervous System. In *Environment and Social Behavior*, Irwin Altman (2) focuses on the function of personal and small group privacy as a mechanism for selective control of access to self or to one's group. In *Design for Diversity*, Greenbie (7) focuses on the other end of the continuum, the need to accommodate the elementary primary group relations, defined as "proxemics," while designing for the large scale cross-cultural associates so necessary in an industrial world, defined as "distemic." Albert Scheflen distinguishes, in *Human Territories: How We Behave in Space-Time* (11), a series of territorial relationships that culminate in built territory. For instance, people may use objects such as furniture to create space oriented to a particular social relationship, and on the next level put fences or walls around those objects, requiring those who wish to enter the group to do so through transitional zones such as corridors or waiting rooms.

Because the thinking in the field of community and privacy clusters both around specific behaviors, e.g., the experience of crowding, and the general issue of scale, we have divided the book into sections which may appear somewhat arbitrary. But we feel that after the general section, a section on crowding and three sections on design of increasing scale address or touch upon most of the current behavior concepts as they relate to design. We trust our classification system will be a means of making the volume more usable and not an attempt to isolate phenomena which are not in reality separable. At the beginning of each section, we will attempt to relate each of the papers to the themes established here and to each other as far as the subject, methods and viewpoints of the authors allow.

REFERENCES

1. Alexander, C. The city as a mechanism for sustaining human contact. In W. E. Ewald, Jr. (Ed.), *Environment for man: The next fifty years*. Bloomington, Indiana: Indiana University Press, 1967.

2. Altman, I. *The environment and social behavior*. Monterey, California: Brooks/Cole, 1975.

3. Chermayeff, S. and Alexander, C. *Community and privacy*. Penguin Books, How Minds Work, 1962.

4. Coss, R. G. The cut-off hypothesis and its relevance to the design of public places. *Man-Environment Systems*, 1973, *3:* 417-440.

5. Esser, A. H. (Ed.). *Behavior and environment: The use of space by animals and men*. New York: Plenum Press, 1971.

6. Goodman, P. and Goodman, P. *Communitas: Means of livelihood and ways of life*. New York: Vintage Books, 1947.

7. Greenbie, B. B. *Design for diversity*. New York: Elsevier, 1976.

8. Hall, E. T. *Beyond culture*. Garden City, New York: Doubleday, 1976.

9. Leyhausen, P. Dominance and territoriality as complemented in mammalian social structure. In A. H. Esser (Ed.), *Behavior and environment: The use of space by animals and men*. New York: Plenum Press, 1971.

10. Lynch, K. *What time is this place?* Cambridge: MIT Press, 1972.

11. Scheflen, A. E. *Human territories: How we behave in space-time*. Englewood Cliffs, New Jersey: Prentice Hall, 1976

Section 1: General

INTRODUCTORY NOTES

The two papers in this section try to define this volume's terminology with an evolutionary and attitudinal inventory analysis. This task is enormous and never-ending, as can be ascertained merely by looking at the divergence of the authors' points of reference. Where Esser searches for complementarity in community and privacy, Turnbull starts out by trying to measure the independence of these terms.

Esser proposes to search for design leading to communality because he believes that within this context community and privacy are optimally safeguarded. He defines community as the feeling of unity created by shared experiences in a group. Privacy can be conceptualized as the feeling of release from the obligations to the community. The need for privacy arises only if individual members of communities discover communal interests with strangers and desire communality, while continuing to need their primary community. We struggle with such concepts as "unity in diversity" precisely because we are searching to accommodate exploration of individually created communal interests in privacy, while simultaneously maintaining the advantages of unity through community participation. Our Central Nervous System creates knowledge of the contexts for community and privacy; that is, our brain tells us when behaviors and places fit to produce feelings of togetherness or separateness.

The designer, in Esser's opinion, should know that creation of context is influenced by synergy, the working together of parts of a system in such a fashion that they increase and potentiate each other. (Synergy can also be negative, as Ruth Benedict (1) described for society.) Knowledge of the role of the designed environment as a code to communicate context is essential for the designer (2). Esser posits that we cannot design for community or privacy as such, but that we can assist the brain in creating contexts in which our needs for these feelings are satisfied. A designed communality would provide places in which behaviors fit and could create a synergic context of mutually reinforcing action-spaces without the loss of community and privacy. Such a design is unavoidably a compromise between possibly divergent directions, but it can also be a liberation from one-sided interpretations of reality and the common good. In this volume, Greenbie's final chapter on the design for a community of diversity refers to a possible solution of communality.

Turnbull's exploration of the relationships between privacy and community uses activity space to arrive at a way in which these three concepts can be tested in an orthogonal analysis as being independent of each other. This offers the advantage of making attitudinal inventory data amenable to statistical manipulation. Turnbull asked ten questions, each along these axes, and the answers from 60 residents of Ottawa and 139 from London are compared. As expected, the British results were overall similar to the Canadian results, but also, as one would anticipate, there were ambiguities. For instance, the relationship between community and activity space in the Canadian sample was significant and positive; in the sample of Londoners, this was not the case. This finding shows the need for further development of inventory tools to be used in research on community and privacy.

Turnbull makes the important point that implicit assumptions about the interrelationships among the concepts studied cannot continue to guide designers. Similar caveats regarding *a priori* positions are raised throughout this volume. It is clear that much more research, both on the conceptual and empirical aspects of community and privacy, is needed.

REFERENCES

1. Benedict, R. Synergy (Some notes selected by A. H. Maslow and J. J. Honigmann). *American Anthropologist*, 1970, *72:* 320-333.

2. DeLong, A. J. Communication as a generic process. *Man-Environment Systems*, 1972, *2:* 263-313.

DESIGNED COMMUNALITY: A SYNERGIC CONTEXT FOR COMMUNITY AND PRIVACY

Aristide H. Esser

The Association for the Study of Man-Environment Relations, Inc.
P. O. Box 57, Orangeburg, New York 10962

ABSTRACT

We ought to design in the image of man evolving, confronted as we are with the divergent results of human creation and their lost contexts. The overwhelming complexity of human endeavors appears to prohibit the synergic benefits of our cumulative experience. I propose to evaluate the potential for synergy of design, based on what is known about brain/behavior/environment.

The Central Nervous System (CNS) as the seat of our experiences, creates context: the fit between behavior and environment. Creation of context is made possible by synergy, whereby parts of a system potentiate each other to provide meaning. Synergy in the brain and in society can be positive or negative, desired or unwanted. In this chapter I shall use the term only in its positive sense as in the creation of context.

The introduction to this chapter will define terminology. The next section offers a discussion of the evolution of the CNS and its environmental counterparts. The important point here is that our environmental realities are perceived differently on the nonverbal and verbal levels of the brain. Additionally, on the latter level, language and human prostheses have created analogues of the functional brain levels in society. These complex analogues appear to lead a life of their own and make it hard for us not to consider them part of the perennial natural world. The following section attempts to explain how the environmental change process responds to recurrent challenges in human evolution. The review of design responses to deal with needs for community and privacy through history leads to a discussion of the reasons for loss of context, especially in industrial urban areas, and why some of the proposed solutions have not worked.

In today's urban environments context is hard to create because the bits and pieces and their functions appear to have lost meaning. A new framework for communality can restore it, and, consequently, a change of paradigm is proposed in the section on obstacles to synergy. The final section of this chapter suggests ways of improving synergy of verbal and nonverbal functions, or the left and right brain hemispheres and their societal analogues, using a complementary model. The chapter ends with a discussion of aspects of synergy useful in design.

INTRODUCTION

Concepts of community and privacy are abstractions of age-old experiences and feelings within certain living situations. A living situation is an interplay between behaviors and environments or a totality of specific human activities and the places in which they occur. This totality is *context*, created by the Central Nervous System (CNS) in combining time and space or behavior and place.[1] The evolution of behaviors and concomitant environmental changes leads to continuous adaptation and creation. These processes rely on feedback and feedforward loops of mutual-causal and interdependent factors in cognition (what the brain creates), behavior and environment. We need a design orientation based on understanding of changes in man-environment relations. This chapter postulates that where human-cognition/behavior/environment systems change little and thus allow for uneventful biological and emotional life, the situation has a *high context* (32). Where relationships change rapidly, *low context* exists, necessitating continual innovation in behavior and environment. In terms of cognition, the CNS creates a high context and perceives a high degree of fit between behavior and environment in an unchanging environment. The opposite occurs in a continually changing environment. For the individual, however, there is always a complementarity between these extremes, in that the CNS tries to optimize context by changing behaviors to fit changes in the environment and vice versa. This search for optimum functioning is the propellent for human evolution.

In evolution, the feeling of *community*, of being at home and satisfied in a familiar sociophysical environment, came first. The

---

[1]Different aspects of the context of built environment, such as the organization of meaning, communication, etc., will not be analyzed in this chapter. A multidisciplinary overview has recently been presented by Rapoport (54).

feeling goes back to a primordial experience of unity. This is because communion with the mother forms the root of community feelings, which we, in common with social animals, transfer to our in-group, the people with whom we share and spend time.

The basic definition of a community would be "a group occupying a territory or area, sharing interests wide enough and complete enough to include their lives" (48). Spatial proximity is necessary for the establishment of involvement. This notion receives support from a sociological concept, the *primary group*, which consists of members having face-to-face contact and thereby evolving *primary bonds* of intimate participation. A family is an example, but any group located in one place and having achieved common goals and identity can be considered a community in the primary sense.

The simplest community is determined by bonds of origin, the primary relationships of birth and place of living--the context of blood and soil. When these relationships are in question, *privacy* is needed. The strange(r) will threaten assumptions of safety or jar us into awareness of characteristics of self which should remain hidden lest we become vulnerable.

In human evolution, *communality* was created when representatives of primary communities met for the purpose of pursuing new interests and goals, establishing unfamiliar patterns of exchange and sharing. Discovering new places and ideas by sharing with the stranger is attractive, as the Bible recounts for Eve: ". . . (The Serpent said) as soon as you eat it, your eyes will be opened and you will be like gods . . ." (Genesis 3, Verse 5, *New English Bible*, Oxford University Press, 1970). Seduction, the promise of sharing another context, is a powerful stimulus for evolution.

Communalities, once formalized, not only became containers for innovative processes, but also providers for familiar contexts in which participants have learned to confront the new. Thus, we can think of communalities as forming the basis for *secondary communities* in which development of such capabilities as sharing, bonding, role-playing and communication is allowed at a safe pace. The increasingly abstract communalities of today may be too short-lived to provide sufficient bases for community as mutual acknowledgement of relationships between behaviors and places. The framework for common understanding may be lost to the participants unless communality is specifically designed to encompass and transform great differences in the make-up and complexities of its constituent communities.[2] In the absence of communal context, the general relatedness of people and space, we look for privacy.

---

[2] See footnote on page 12

With the Fall, because of the Serpent, the unfamiliar other, humans experienced for the first time a discontinuity in the communion between Nature, Self and God. Innovation took place: "Then the eyes of both of them were open and they discovered that they were naked; so they stitched fig leaves together and made themselves loincloths . . . and hid . . ." (Genesis 3, Verse 7, 8, *New English Bible*, Oxford University Press, 1970). This episode shows human construction of a new context. Private behavior (to be inconspicuous in public) and changing the environment (to provide cover) occur together and become *privacy*. When feelings of community break down or do not apply in a new situation, when a general context is lost, we can no longer share meaning with everyone.[3] At such times, one needs to allow for an appraisal of what happened and for exploration of the fullest range of emotions through appropriate behavior (e.g., intimate sharing) or environment (e.g., a retreat).

Privacy may be defined as relationships (47), implicitly or explicitly consented to, which indirectly influence environmental design. But in this age of information exchange, privacy, for practical purposes, is mostly defined as the claim to self-determination of when, how and to what extent information is communicated (60). In that case, there is a direct influence on environmental design since this clearly can control information flow (1) and make one feel secure to:

1. Preserve personal autonomy, thereby fostering the growth of individuality,
2. Afford emotional release,
3. Provide for self-evaluation,
4. Enable protected communication,
5. Allow transition from one state to another.

---

[2]For instance, Maruyama suggests that if we want a cross section of the world population and its skills to serve the communal goal of extraterrestrial colonization, we may need to experiment simultaneously with such opposite and complementary types of communities as the hierarchical homogeneous community which seeks to maximize the "best for all" and its majority rules, or the symbiotic heterogeneous community which seeks to choose and match individual concerns without attempting to maximize "the best" (46).

[3]Stimulation without meaning results in a feeling of overload, which explains why experiencing crowding or being unable to "have one's way" leads to such attempts to secure context as escape behavior or change of environment.

In other words, individuals who are to function in evolving communal interests need contexts in which exist appropriate environmental safeguards for new types of behavior.

Experiences of communality and privacy resulted from the voluntary break with community to pursue self-realization. Jung described this process of individuation as a complement to being part of the community and having access to the "collective psyche" (37). It is part of the step to man and characterizes the constant search to "better" itself, which differentiates the human from the animal brain. What can we learn if we conceive of our CNS both as instigator and consequence of processes by which we adapt, create and adapt again to our creations?

EVOLUTION OF BRAIN AND BEHAVIOR

We may view the organization of behavior and environment as a reflection of the organization of our CNS, which, according to MacLean's concept of the evolution of a triune brain, contains three different interlocking subsystems, each with its own mentality (43). These CNS subsystems create contexts on different levels of functioning. But most of the time, we are unconscious of possible disparities since synergy between the subsystems continuously provides for construction of that one reality which is our experienced context. Table 1 summarizes the evolution of contexts.

The basic neural machinery of our brainstem was the first to evolve, driven by what MacLean calls the "reptilian forebrain" (43, p. 216). Most reptiles appear to function in physical space only. Actions are prototypical for survival and often stereotyped according to the place where they occur: establishment, marking and defense of territory; homing; ritualistic displays in aggression and sex; etc. This CNS subsystem is intent on maximizing individual security and shows only primitive social behaviors, based on mutually acknowledged ownership and procreation. I have called our understanding of the behavior patterns basic to the survival of members of one species the context of the *Biological Brain* (20). Its instinctive vital behaviors and the resulting exclusive relationships become more complex and malleable with the evolution of the second CNS subsystem which introduces learning.

The limbic system, or the lower mammalian brain (43), makes possible a transformation from physical to social space. Because the mammal depends on its parents for survival, social bonds surpass the security of physical place. Life events are experienced in terms of pleasure/displeasure. This emotional mechanism allows memorizing and makes learning and ordered group life possible. Mammals apparently function in social space; they know their rank

TABLE 1

HUMAN CENTRAL NERVOUS SYSTEM AND EVOLUTION

CENTRAL NERVOUS SYSTEM			BEHAVIOR		ENVIRONMENT	
Level	Subsystem	Cognitive Function	Action Space	Relationship	Environmental "Brain"	Typical Structure
Higher Mammalian	(Neo)cortex* (verbal)	Intellectual	Conceptual	Complementary (role)	Prosthetic	Interaction Territory
Lower Mammalian	Limbic System (nonverbal)	Emotional	Social	Reciprocal (status)	Social	Dominance Hierarchy
Reptilian	Brainstem (nonverbal)	Vital	Physical	Exclusive (ownership)	Biological	Territory

*I am aware of possible transformations of behavior and environment if the right hemisphere is conceived as a projection of subcortical systems (26). Further discussion of left and right cortical functions will follow in the last section of this chapter.

relative to group members and establish leadership and preference in addition to the retained "reptilian imperative" for physical security. Ownership, the expression of exclusive relationships in physical space, transforms into status as an expression of reciprocal relationships in social space. For instance, instead of controlling specific objects in space, the dominant animal may exert social control, having preferential access to food or first right privileges to mount receptive females. We know that the human struggle for power is often not for possession but for status or position, the power of decision. The focus of control thus shifts from space to time and is overtly manifested by a working concensus whereby group members acknowledge each other's rights and duties--a dominance hierarchy. A context of hierarchy, combined with other manifestations of social regulation in a species, may be called a *Social Brain* (20). The hierarchy based on reciprocal relationships in small primary groups provided an advantage for survival, since the established social rank of each member reduces intragroup competition. Not only is energy saved, but the internal group discipline is useful when intergroup fights occur or communal purpose and intent has to be served. Individuals with the ability to differentiate clearly between group members' strengths and weaknesses would be able to make the best decisions in combining subgroups formed for such activities as hunting, foraging, warfare, etc. But in large groups, concrete memories of positive and negative experiences with other individuals can no longer guarantee choice of optimal behavior. Generalization of individual characteristics accelerated the pursuit of communal goals with many others or strangers; thus the tool of general symbols (language) and reasoning evolved to overcome the limitations of limbic functioning. The cortex became man's prosthesis, the abstract extension enabling us to become (self)conscious. In other words, vital and emotional functions are nonverbal and could not come under conscious control until the higher mammalian or third brain level evolved as a tool for the expression of our earliest and basic experiences. Seen from this perspective, cortical functions protect and shelter our emotional life, as does keeping a poker face.

Language, memory, logic and reason made possible the incredible range of human behaviors and artifacts we witness today. Planning and design, the pretesting of new behaviors and new environments without possible disastrous biological and emotional consequences in real life, became the rule when the cortex created conceptual, nonphysical space in which we interact as soon as we learn manners and language. The indirect role of the cortex in dealing with the environment is also evident in its description as the conceptual brain extending the capabilities of what Greenbie calls the dracolimbic brain--brainstem plus limbic system (31)--and in Gray's theory that knowledge is organized according to emotional nuances (30). Ultimately, all decision making is limbic; if knowledge is limited, we "act on gut feelings."

The advantages of mastering knowledge is that it gives one the ability to participate in many interaction territories. This enlarges one's alliances, since one's value to a particular group would rest on one's contribution to its action space. Recognition among group members of potential complementarity of functions leads to definition, acceptance and learning of roles. This capability reinforces the exclusive and reciprocal relationships mentioned earlier and thus permits growth of complex communalities in which individuals may participate in many communities by playing many roles.[4] Plurality of contexts is inherent on the third brain level, and the collectively agreed-upon connections between aspects of our man-made environment and our conceptual spaces may be called our *Prosthetic Brains*.

Prosthetic brains proliferate themselves as environmental embodiments of perceived communalities. These, in turn, can be conceived to exist on similar functional levels as the brainstem, limbic system and neocortex. Table 2 illustrates analogies between structure and function in CNS and society. The vital, biological functions of our brainstem, keeping us alive, are represented in our economic infrastructure, the communalities of commerce and industry. The emotional, social bonding functions of our limbic system are reflected in policy, legislation and decision-making functions. Acculturation is possible when government enforces education as a counterpart to learning within family and group. This brings us to the level of the neocortex, where creative communications cause ideation and cultural life: religion, the arts, sciences and technologies. The interplay between all of these communalities determines our individual behavior.

We create contexts by any combination of behavior and place in the biological, social and prosthetic brains. But these individual creations must fit conceptually with group or societal contexts created by the communal interests of economy, government, religious or cultural life. The individual can only temporarily pursue a context discongruent with that of others without his communication seeming alien to his fellowmen. In that case, he will run the risk of being considered inappropriate, a madman, or at best, an impractical genius. In other chapters of this volume, DeLong and Greenbie discuss design implications of the fact that context is created with inputs from all three CNS levels, not only from the cortex or what Greenbie calls the "planning brain" (31). One can also say that creation of context, in addition to being a socio-logical event, is also a neuro-logical event, following the

---

[4]This goes for groups with communal interests as well, e.g., ". . . members of the Hyde Park-Kenwood Neighborhood Council attribute part of their success to having learned of and established contact with over 130 municipal bureaucracies" (57, p. 11).

TABLE 2

SOCIETAL ANALOGUES TO CNS STRUCTURES AND FUNCTIONS

CNS		SOCIETY	
Subsystems	Functions	Structures	Functions
NEOCORTEX	Conceptual/ Intellectual	Arts, Sciences, Technologies, Religions, etc.	Ideation, Cultural and Spiritual Life
LIMBIC SYSTEM	Social/Emotional	Governance	Policy-setting, Decision-making, Legislation
BRAINSTEM	Vital/Biological (Basic neural machinery)	Commerce, Industry (Infrastructure)	Economic survival, Enabling enterprise

logic of evolved CNS structures. This explains why communality (and privacy) can be discussed in terms of relationships (47) or relational logic (DeLong, this volume).

The three Brains and their societal counterparts ideally are synergic for mankind, collectively or in any possible combination. Thus, in gauging the potential for positive change in any situation, one has to look for potential synergy between components of brain, behavior and environment, their different levels (Table 1) and the conceptual communalities (Table 2). The simultaneous interaction between all levels of man-environment systems creates a weave of mental and environmental actions. Furthermore, the creative impulse originates from our oscillation between animal motivations and human goals.

Man, as an ape, freed his hands by virtue of adapting the upright posture. These hands learned to work, to improve and change the natural environment drastically. And the creations of culture set man apart. These conscious transformations trigger the individual to strive for well-being by innovation or by conceiving of quality. But being well is being whole. And this goal, in turn, expresses that which we lost as soon as we began to make ourselves: the sense of wholeness, of being at one with others and with the universe. We have to transcend ourselves in order to be, and this is the dialectic process between the individual and mankind, between privacy and community.

EVOLUTION AND ENVIRONMENTAL DEVELOPMENT

We have seen that changes in our behavior and environment can be related to CNS structures and functions. Evolutionary stages can be considered times during which responses to novel situations are being tested. Challenges first interact stressfully with the old CNS structures and functions, but when new behaviors have been elicited and found appropriate, they are integrated into the former ways of doing things. New CNS structures then function to create new contexts and awareness of how novel situations can be matched with certain behaviors. Because of this cyclical process, we understand when to do what in more and more different circumstances.

Our animal ancestors had nonverbal brains, therefore matching new behaviors to a changing environment was a chance event; fossils tell us about absences of environment/behavior fit. With early man, the breakdown of context in the natural environment occurred so slowly that the neocortex could try out and match behaviors at a relatively slow pace. For modern man, situations often change too quickly to be able to try out behavior. Instead, when context breaks down for us, we have perceived the task to be additional design, often specialized to adapt to needs for existing behaviors. We thought of compromise, such as the addition of recreation environments to offset the tedium of production-line behavior. Now we have begun to understand not only that unlimited accommodation to the man-made world may be dangerous in itself, but that we are better off when environmental design brings about an expansion of our range of behaviors. We will examine the synergic requirements for what may be called the empathic, forward-looking movement in design (25) later on. First, we must look at historical cycles in the design for community and privacy to understand continuous transformation of human contexts and the danger of the specialization it engenders.

Change in our imagined or real environments will never end. The Prosthetic Brains pull mankind into the direction of change as a result of the need for more sharing in a diversifying world. Cultural man set this runaway process in motion by using the tools of the neocortex, a tale similar to that of the Sorcerer's Apprentice (23). The need for familiar high context situations as a refuge from this stress is maintained by the nonverbal brain systems which guard against too much change that might endanger survival (Figure 1). Mankind's present hope for continued evolution lies in the frontal lobe of the cortex, which contains the humane or empathic potential needed to promote synergy between the Biological, Social and Prosthetic Brains (9, 21). The forward and backward pull in the CNS caused built-in antagonisms, as well as mechanisms for conflict resolution, in order to enable action in

Figure 1:
Complementary and Oppositely-Pulling
Subsystems of the Human CNS

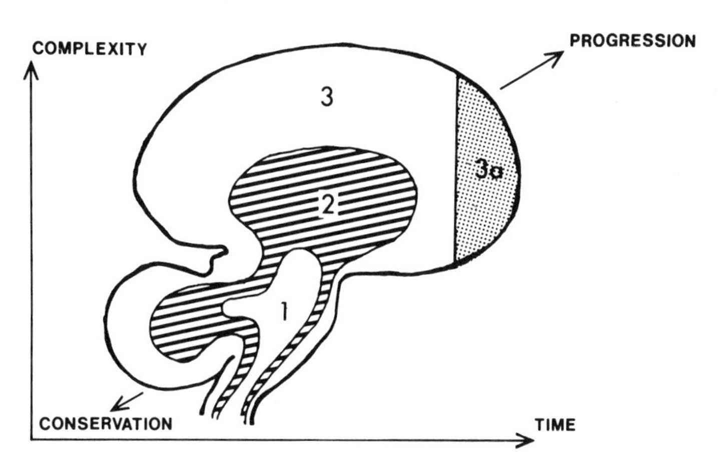

1 Reptilian (brainstem)
2 Lower mammalian (limbic system)
3 Higher mammalian (neocortex)
3a Humane (frontal lobe of neocortex)

Figure 2:
Integration Stages of Successive CNS Subsystems

STAGES	BRAINS			AMOUNT OF SYNERGY
HUMANE	BIOSOCIAL		PROSTHETIC	FULL: Empathy for reason/emotion/ instinct
		*		
HIGHER MAMMALS/ MAN	BIOLOGICAL	SOCIAL	PROSTHETIC (Cortex)	PARTIAL: At times reason-emotion clash
	*			
LOWER MAMMALS	BIOLOGICAL	SOCIAL (Limbic System)		PARTIAL: At times emotion-instinct clash
REPTILES	BIOLOGICAL (Brainstem)			GOOD: Instinctual life

*Potential conflict

Figure 3:
Developmental and Evolutionary Cone

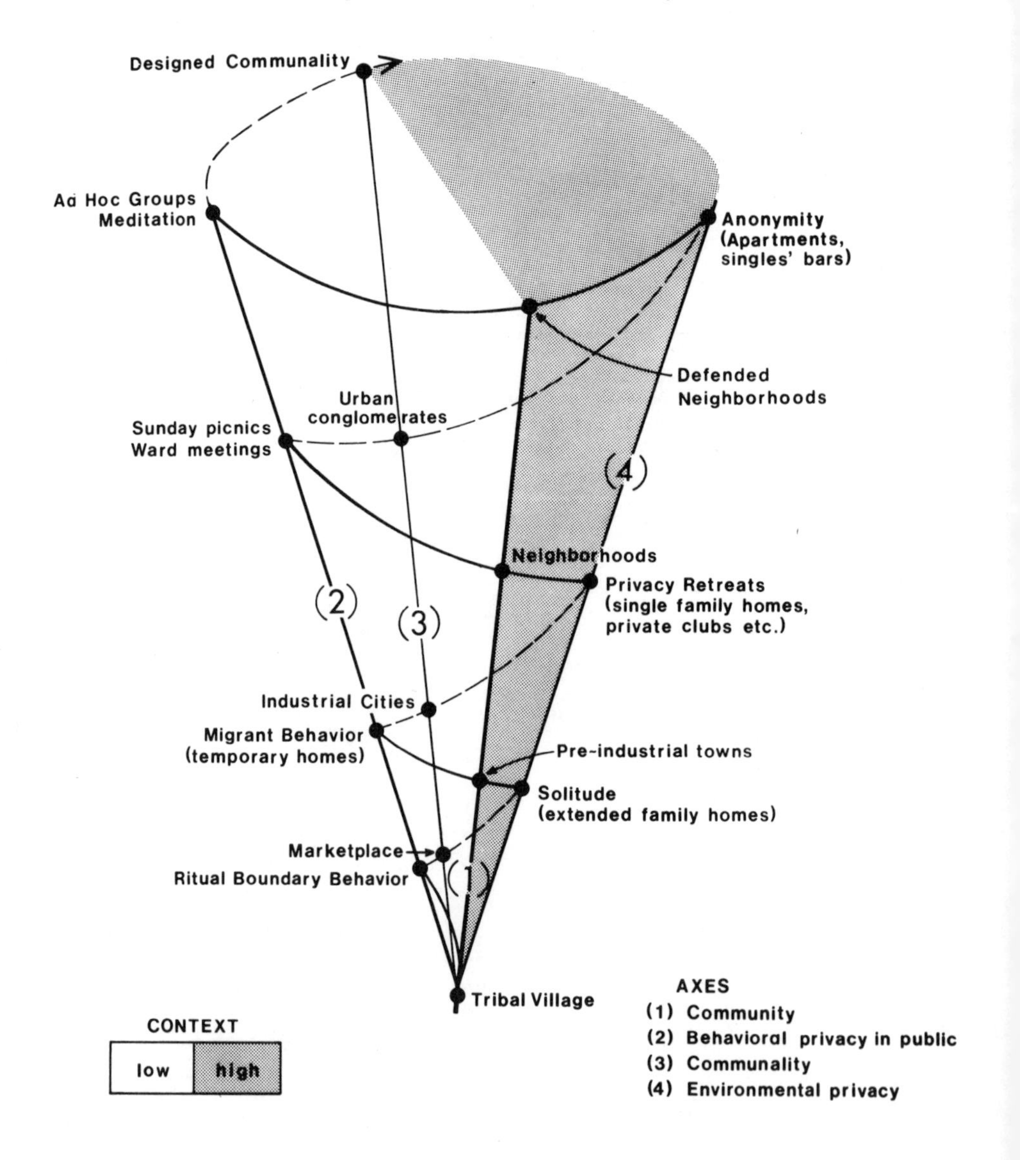

increasingly complex behavior/environment interplays. For instance, in mammalian behavior there is a contrast between instinctive stereotype (brainstem) and emotional response (limbic system). A crocodile would not hesitate upon encountering a stranger, but a dog may simultaneously bark and wag its tail in this circumstance. Similarly, in man, conflicts between emotion and reason are proverbial, and our inclination to either/or behavior often prevents our being humane. Only when we understand that there is room for both, the principle of complementarity, can there be full synergy between all CNS subsystems (Figure 2). And since the basic systems in the brain are built in, there will be no one permanent solution to our problems of living.

In Figure 3, I suggest the image of a spiraling staircase upward a broadening cone of evolution, whereby at each full turn mankind is confronted with the same man-environment relations problem, but on a higher level of complexity. There is then a progression in expressions of (primary) community, followed by behavioral privacy, then communality, leading to environmental privacy, a recurrence of (secondary) community, and so on. Sometimes this happens simultaneously, as in a multicultural locality. Most of the time there are successive developmental stages.[5]

One possible developmental construction assumes that in the history of the tribal settlement there must come a moment whereby the context becomes so high (the relationships between behaviors and places are in harmony, as in certain Balinese villages) that innovation and change is precluded. When tribal members feel the creative urge under these circumstances, they must look for environmental innovation since behavioral aberration is personally more dangerous than facing the unknown. This may lead to spatial exploration and encounter with other tribes. At such moments, the need arises for behavior with strangers in non-community, or what we may call behavior in public. And the need for privacy (control of information exchange) under those conditions leads to making oneself known with restrained behavior--blending into the landscape, being inconspicuous and least threatening to the other. An example of this behavioral privacy in public is the ritual crossing of territorial boundaries among Australian aborigines. The visitor lights a fire at the boundary and then waits until he is noticed and someone from the other tribe comes to fetch him for entry.[6]

---

[5]This sequence is also revealed in the history of the commune movement, where participants constantly veered between communal space and private quarters (34, pp. 329-335).

[6]Similar boundary behavior can be seen when a new arrival tries to join a group at a cocktail party.

When individuals and tribes have sufficient permanent interests in common, their communality may lead to creation of the marketplace, where historically strangers can meet under mutually agreed-upon rules.

In reviewing this development from the design perspective, we observe that from a prototype of behavioral and environmental unity (in the high context village, all relationships are stable and design may not require conscious effort) design for diversity must occur. As soon as contacts with strangers become regular, the need arises to meet formally for purposes of exchange in a specific place. The need arises also to provide for behavior conformity in a deliberate manner, agreement as to where and when to speak on what topics, what language to use, etc. In other words, *design decisions have to be conscious when context has broken down.*

But in a marketplace, communal enterprise unavoidably increases the chances for unexpected threats to individuals and families. Individuals will have to look for environmental privacy because showing privacy behaviors in a very diverse public place could be misunderstood and lead instead to exposure. Apart from that, contact with strangers of different background and with different skills strengthens individuation. Individuals begin to regard themselves and their families as unique, and this, in turn, quickens the need to secure food, space, etc. Additionally, the experience of crowding, when one feels unable to have one's way, begins to surface and requires some sort of stimulus control or control of space (2, 7, 47). Consequently, in contrast to the free and practically unencumbered sharing of information and responses which is the essence of community, restriction of physical accessibility is sought, in solitude or with a few intimates.

Through the ages, in different cultures specific design solutions have been consciously attempted to provide for privacy. A first step might have been to safeguard privacy by providing for solitude or building a home for the extended family, rather than live in a tribal home. Within a pattern of family homes, possibly of different clans, together with structures enabling storage and exchange, one can begin to feel safe again. Context may return, and with it a feeling of community in addition to that of privacy. This may have been the characteristic of life in a city-state or pre-industrial town. It is only to be expected that growth in awareness and skills in generations of townspeople under favorable conditions would recreate a high context community, more complex but otherwise similar to a tribal community.

Again, history has shown that the development of industry destroyed context, notably that of small pre-industrial towns. The changeover to industrial conditions created migratory behavior, and the industrial city allowed expressions of behavioral privacy by setting up inexpensive housing or temporary homes in certain spaces. Accommodation of communal interests brought about low context; one need only read Upton Sinclair's *The Jungle* to perceive the uncertainty as to how behavior and environment related in such surroundings. For a child growing up in the deliberately chaotic (to ensure lack of identity and thus cheap labor) context of the Chicago stockyards, feelings of community and privacy came few and far between.

In the industrial town there has been a marked expansion of interest in the family at the expense of public life (4). One-family homes for the nuclear family which now cannot afford to have unproductive grandparents around, clubhouses, corner bars or grocers, union headquarters, etc., were environmental expressions of privacy, as were the merchants' quarters in the marketplace. Privacy retreats, adaptation to threats to self, in turn might form the nucleus of a perceived neighborhood through which community feelings were reconstituted. Neighborhoods and their organizations were for families who had migrated in search for work what the bonds of blood and soil had been for tribesmen. Once life in the neighborhoods of industrial towns had lasted long enough, the communities would be quite homogeneous and offer high context. For instance, even after World War II, some parts of early industrialized Leyden in the Netherlands, were still exclusively related to the specific textile industries established over the last few centuries. I knew inhabitants of these wards who would not dream of a different career and marriage or travel outside of the city.

The high context of neighborhood faded with the arrival of readily available means of transportation, especially the automobile. In a mobile society, the public rituals of Sunday outings or closed meetings for political, religious or other purposes expressed design for privacy. This heralded modern times in which communality progressed by conglomeration rather than integration. We appear to be unable to make a coherent whole out of the infinite multitude of common interests as we move about in post-industrial society.

Whether we are an industrial or post-industrial world, the number of people living in cities will grow phenomenally. Urban growth is well researched in the Chicago school of "urban ecology," describing expansion of the city in circular fashion. The center became surrounded by a transitional (often poverty) area of people with menial jobs. Around that grew rings of apartments and one-family dwellings, merging with suburbs of better economic status.[7]

---

[7]This process took place mostly in the U.S., far less in more homogeneous European cities.

The confluence of suburban settings led to the present pattern of conurbations or metropolitan areas whose evolution has given rise to concerns over the well-being of large parts of their population. The context of this latest form of communality, while furthering privacy needs, may be too low for the development of viable communities.

Before the reasons for and the implications of this statement are discussed, let us briefly outline some of the (possibly counter-productive) expressions of further specialization of privacy and community design in urban conglomerates. First was the search for anonymity, in apartment living, etc. In metropolitan areas, anonymity is a form of privacy. Much of design for anonymity has been brought about inadvertently by the sheer mass of urban building, but conscious design for anonymity is developing.[8] Then some feeling for community was obtained from what has been called "defended neighborhoods" (57), where exclusive behaviors (restrictive zoning, vigilante patrols and other ad-hoc reactions) and territorial expressions remain the only substitutes for the loss of primary social bonds. Although the defended neighborhood may be viewed as a creative response, especially in the inner city poverty zones, the term reveals its essentially conservative function, which can lead to stereotyped behavior, counterproductive for the city as a whole.

The unavoidable contact that even the most private person must have in a world that evolves through communality makes it attractive to design more and more private environments and develop new forms of privacy to use in public. Certain reserved behavior patterns which either indicate the need for very partial sharing (ad-hoc special interest groups) or "advertise" a need to join with unknown others sharing one's special interest (e.g., when meditating in public) have become acceptable in our metropolitan areas. In light of the original feeling of community, anonymous central core urban buildings and anonymous behaviors are absurd solutions to regaining feelings of privacy, but their increasing occurrence indicates that prevalent economic reasoning and conurbation preclude other solutions.

At this point we may be reminded that previously the boundary between private/public was clear, as in the Graeco-Roman world where the public realm was defined as an area of freedom and equality (51). But with the industrial revolution, privacy has increasingly become associated with certain exaggerated aspects of the individual. As

---

[8]For instance, anonymous exteriors and location of mental health facilities have been proposed because patients want to hide their visits (24). The conscious design decision is a step forward for the elitist mental health profession, but it does reveal lack of community.

Nisbet observed, ". . . individual is used in different contexts. Today it is an abstraction within an institutional context, where the attribution of properties of our society to the interior life of the individual have produced the image of possessive individualism" (50, Ch. 10). Once privacy is tied to the idea of the individual being the proprietor of his or her own person, society preponderantly becomes a network of exchange relations between people on the basis of ownership. This, as we saw in the discussion of our CNS, is a reflection of the simplest level of organization. Exclusive attention to the Biological Brain and territoriality leads to excessive valuation of economic infrastructure. On this basis, we may diagnose the schizophrenia in our design solutions. As O'Neill states, "All the spaces of the modern world are absorbed into a single economy whose rhythms are linear and mechanical. The architecture of public and commercial institutions, the furnishings of the house, and even the styles in which we clothe our bodies threaten to destroy the dialectic between the things that are to be shown and the things that are to be hidden. The results vary from the inhuman naked space of the typing pool to the democratic open spaces of Toronto's new City Hall where the shocking exposure of secretarial knees produced demands for privacy in the design of working areas" (51).

The spiraling road climbing around the evolutionary cone symbolizes the broadening of issues as they become more complex and the inevitable divergence of our solutions. At any given time, a cross section of the cone would provide examples of expressions originating from a search for community. These expressions can be grouped around opposing axes: community and communality, behavioral privacy and environmental privacy. Thus, in a cross section of the typical industrial city we may find migrant behavior, single-family homes and neighborhoods. Following its axis as a record of environmental history, the industrial city may show remnants of the old marketplace next to beginnings of suburbia, maybe even a project attempting designed communality.

The above brief excursion through the history of a complex evolution highlights abstract milestones of a recurring process in the development of community and privacy. For instance, anonymous people in the metropolis may make up totally new areas of communality to obtain the satisfaction of face-to-face contacts lost with their primary groups. Singles bars or bathhouses are a good example of the search for biosocial need fulfillment without the territorial image of community. For such bars, non-distinct environmental design is functional, since the visitor does not seek environmental evocation of place but wants shared informal behaviors. But this example already illustrates a distorted context, revealing the basic incompatibility of our behaviors and the (non)design of modern (sub)urban areas.

INDUSTRIAL URBAN CONDITIONS AND IRRETRIEVABLE LOSS OF CONTEXT

Since it appears inevitable that by the year 2000 more than half of mankind will live in cities, working on design for urban context is the great challenge (54). The social sciences traditionally linked industrialization and urbanization to a general decline in the integration of society or decreasing context. A decline in social cohesion creates special design problems. For instance, diminished functioning of the family creates institutionalization, resulting in segregation of such "special" populations as the elderly and the mentally ill and bureaucratization of the work setting. But as we have seen, making more and more special contexts leads to a widening vicious circle of less and less compatible design responses to needs for community and privacy. Additionally, because urbanization leads to the concentration of people and industry, it is potentially upsetting to the ecological balance, threatening exhaustion of natural resources and the quality of the environment.

The likelihood of the absence of communal goals increases for the population in comtemporary metropolitan areas. In the past, new developments and activities appear to have temporarily shaken up established contexts to allow emergence of a larger framework. With conurbation, this has not occurred. Instead, the socio-territorial polarization of poor inner cities, often with minority populations, and rich suburbs preclude further integration of and cooperation with each other's interests. The development of territorial (ethnic) communities or defended neighborhoods are no advance either for the metropolis or the suburbs because they may lead to permanent antagonism between factions.[9] Revitalization efforts of central districts, which have retained their attractiveness to business and trade, are doomed if the surrounding zones of poverty remain socially polluted and crime-infested and continually use a disproportionate part of the city's economic base for services. This vicious circle of inner city renewal and erosion of its profits through decay of slums is fueled by the increasing unemployment of the slum population.

The inner city worker often commutes by car from the rich suburb, but the city dweller often cannot serve the high technology needs of or commute to jobs in the suburbs. (In the last decades more than 50% of manufacturing jobs originated in the city have shifted to the suburbs.) Restrictive zoning laws in the suburbs

---

[9]A projection of defended suburbs around golf links in combat with inner city marauders in the 80's has been provided by Walker Percy in *Love in the Ruins*, Dell Paperbacks, 1972.

do not make a permanent move for most city dwellers feasible. Less and less of those trapped in the cities are employable. And as the Joint Economic Committee of our Congress states on the basis of the Brenner Report: ". . . this study reveals that unemployment has a strikingly potent impact on society. Even a one-percent increase in unemployment, for example, creates a legacy of stress, of aggression and of illness affecting society long into the future" (8, p. V). A look at some of the indicators of social stress which traditionally have been related to industrial society's "anomie," or what I would call loss of context, reveals that a 1% increase in unemployment leads to increases of 5.7% in homicide, 4.1% in suicide, 4.0% in state prison admissions, and 3.4% in state mental hospital admissions (8). I am saying that the growth of conurbation did not in itself lead to unemployment, but to a negative synergy of many socio-economic factors. Socio-territorial polarization has made large urban areas vulnerable to increases in national unemployment and increasingly less likely to profit from an upturn in the national economy. Thus, conflicts of interest between parties in the metropolis become more firmly established, and reconstituting context becomes more impossible day by day.

It is a combination of factors in urban life and technological development that drives us further apart, not industry or cities by themselves. Competitive mechanisms that are the tradition of industrialization make synergy in large-scale living situations unlikely.

Such insights are not new, and several deliberate solutions for this dilemma of built-in conflicts of interests and the unavoidable filters in our perception have been proposed. The oldest attempt is the *construction of Utopia*, in which everything that is known to be in the best interests of mankind is harmoniously brought together. The inherent shortcoming of this solution is that it cannot take evolution into account; as we know since the last century, evolution is unpredictable. Perhaps the *growth of totalitarian solutions* is a consequence of this discovery. The motto might be: "If it cannot be predicted where mankind will go, let us make sure to remove its freedom of choice." There is no reason to fault this solution *per se*, but we reject it because of our choice of the principle of self-determination, which includes the right to privacy.

Experimental communes have been formed, trying to incorporate individual freedoms while maintaining the process of sharing (the principle of communality). Some types of communes successful on a small scale are the self-sufficient agrarian settlement, the extended family community and the ideological support groups. They are based on the idea of material and social equality, justice, common property of income and of production. To safeguard self-determination, most contemporary communes hold the individual in high esteem and stress individual responsibility. In contrast to some communes of the previous century based on new industry, science and technology (34), contemporary communes are often pervaded by ecological concerns and try not to use more than what Schumacher calls "intermediate technology" (56). A commune is attractive as an alternative to low context conurbations because it is designed for desired behaviors. However, the commune movement, important as it is in providing mankind with new experimental insights, cannot satisfy the raised expectations of the earth's billions of people, who experience bits and pieces of human communalities through the media.

If a Utopia, a totalitarian state or experimental communes cannot provide the evolutionary context for all of mankind, perhaps a combination of several existing or proposed solutions will. The optimistic approach of designing for ethnic or cultural homogeneity was proposed, especially in younger nations. But as is clear from current dissatisfaction with the concepts of society as a melting pot for, or a mosaic of, cultural identities, we need to think of designing a multicultural society (e.g., 36, 61). I believe that such proposed solutions as open-ended environmental design (52, p. 364; 54, p. 355) are a necessary, but not sufficient, contribution toward the potential accommodation of cultural pluralism.

There is also the approach advocating a simpler life, be this on a smaller scale or with simpler technologies. The smaller scale by itself would be no solution. As Bell states: ". . . regionalism would provide no real solution, for the definition of a region is not hard and fast, but varies with different functions: a water region, a transport region, an educational region have different overlap on the map" (10, p. 45). In other words, as we have discussed repeatedly, the complexity of modern life forces development and growth of communal interests on increasingly larger scales. This is also the reason why urbanization as an ecological niche for modern man offers something unique. Its scale guarantees a favorable nutrient medium for (1) the accumulation of the most advanced cultural concepts, (2) the intensification of human contacts and social intercourse, and (3) the production of information needed for the comprehensive growth and development of the individual. However, even if we cannot turn back from urbanization, we can diminish our present reliance on complex industrial technology (e.g., 28). It is not within the context of this chapter to expound the merits of this approach,[10] and I will only refer to E.F. Schumacher as its

---

[10]For cumulative argumentation read, for instance: *Environment*, 500 Trinity Avenue, St. Louis, Missouri 63130; *RAIN*, Journal of Appropriate Technology, 2270 N.W. Irving, Portland, Oregon 97210; *The Ecologist*, Journal of the Post-Industrial Age, 73 Molesworth Street, Wadebridge, Cornwall PL27 7DS, U.K.

leading advocate. (56).

The above makes the point that reconstitution of context cannot be brought about with any one of past and present approaches. Our contemporary situation is one of muddling through or making do --apparently common sense in a complex world struggling to define what all people have in common and what will make mankind's survival possible. But although it is necessary, it is not sufficient to make for quality of life. The approach is based on knowledge of our past and its expectations, which, as we have seen before, is very much determined by man's animal brains. If this approach takes the human part of our CNS into account, it is primarily with a view toward certain functions of our highest brain level, especially science and technology. But one rarely hears of empathy or the frontal lobe in relation to qualities that pull mankind forward. Why?

Let us for a moment return to Table 2 and Figure 1 and try to describe the conservative and progressive functions of the CNS in terms of roles in society. Societal structures are part of our conceptual space, but as we said before, the Prosthetic Brains which create it have functional similarities to the human CNS. Society's biological brain (especially business) and social brain (governance, law and education) are the real decision-makers, and they like to do it on the basis of precedent. That is why one hears of no difficulty when science and technology are proposed to deal in established ways with existing problems, but there are objections to innovative action on the basis of anticipated problems perceived by empathy.

Frontal lobe empathy (21, 42) and, concomitantly, development of an empathic societal function will, in my opinion, lead to a further synergy in evolution. Representatives of the empathic societal function are not readily recognized as a separate group, vocation or profession. Rather than defining them, I can give some examples which show potentially synergic relationships between representatives of different prosthetic brain functions. When Maslow proposed self-actualization as one of the necessary motivations for modern man, managers, if they paid attention at all, warned that following this principle in the workplace would lead to chaos and certainly lessen profit. But Maslow's theory of a hierarchy of human needs fits the concept of synergy and is now part of the human potential movement. It is as well becoming accepted in industrial design (38) and management (3, p. 157-164). Buckminster Fuller offered countless synergic design solutions (27) which have been ignored by industry or investors, even though some of his ideas have great economic potential. But the goals and format of synergic ideas are often antagonistic to the territorial and dominance patterns existing in traditional economic infrastructures and politics. In the U.S.A. only recently has government supported future planning which makes reference to energy saving, doing more with less and cooperation between all factions of society, including industry and public interests.[11] Is this the beginning of synergy and can it provide context?

OBSTACLES TO SYNERGY AND A PARADIGM CHANGE

We know that the individual human brain, notwithstanding built-in antagonisms between functions of its subsystems or social pollution (20), is synergic most of the time. This means that the functions of its subsystems increase and potentiate each other so that the meaning of the whole exceeds that of the sum of its parts and new and unexpected results may occur. Synergy, therefore, has the aspects of wholeness, harmony between verbal and nonverbal CNS systems or between reason and emotion, self-awareness as well as willingness to act each other's roles and thereby allows us to direct our evolution (23). Synergy rests on empathy, the ability to *identify* with the needs (brainstem level), the feelings (limbic level) and the thoughts (cortical level) of others. In terms of Figure 1, this implies subordination of all CNS subsystems to the frontal lobe because its empathic potential is best equipped for planning (21, 42).

Empathy also provides the potential for synergy in society. To quote Ruth Benedict, ". . . I shall speak of . . . cultures with high synergy where (the social structure) provides for acts which are mutually reinforcing . . ." (44, p. 156). In such societies, if and when there is a pursuit of a greater goal, members become mutually subservient.[12] Unfortunately, there are few inspiring historical examples of synergy besides war. But we do know that human activities are not externally determined, as proposed by Hume and elaborated upon by the logical positivists of the first half of this century. Instead, we believe more and more in our capacity to change ourselves and a new paradigm by which to live. One of its characteristics is a suspension of disbelief based on endurance needed for learning new behaviors and living in changing environments. Once we experience this, it is possible to become "self-conscious" actors and purposefully change our destiny. We then know that people of good will can plan together in new communalities which,

---

[11] e.g., The Committee for the Future and World Future Society, both in Washington, D. C., and Futures Conditional, Spokane, Washington.

[12] The motivation for this type of behavior has become an important topic for psychological study, as in research of altruism.

even if disbanded later, prove that incredible accomplishments are possible.

Can we systematize our knowledge of what is blocking synergy in design and thus preventing context? Built-in conflicts defined by the brain/behavior/environment model, certain aspects of our modern industrial world, and rooted psychological obstacles to understanding are interrelated forces which must be transcended if we want to experience well-being and create wholeness.

According to the triune brain model, we can expect human motivation to gallop off in different and often antagonistic directions. On the biological brain level we are motivated by survival, exemplified by our clinging to territory, turf or possessions. An unempathic planner, unwittingly threatening territoriality, is responsible when individuals enlarge the perceived real or imagined threat into consciousness. Then, any empathy for the planning goal is replaced by antagonism. If not resolved, this will bring about deep-seated and possibly irrational fears leading to passive-aggressive acts and doom of the project.

On the social brain level, there are allegiances to relatives and friends. Individuals who feel obligated will find it difficult to have empathy with alternate courses of action. This explains why rationality is always dependent on the social context. What is reasonable for one may be nonrational for the other. We try to use money as the equalizer to smooth all ruffled feathers and mend injured bonds. But money alone cannot buy synergy.

Finally, on the prosthetic brain level, there are commitments to concepts, especially those grounded in science and technology. And, as discussed in the section on brain and evolution, such commitments ultimately rest on emotional choice. Therefore, chances are that in our actions on behalf of appealing ideas there is little empathy for outsiders (e.g., other cultures), witness high technology proposals for the third world. In our society, high technology has entrenched ways of thinking like "Bigger is Better" and the reliance on weaponry and energy to safeguard our way of life. This does not lead to insight into the needs, feelings and concepts of others, necessary for the planning of alternatives. Yet, we know that from the point of view of global planning, maximizing our selfish interests, maintaining a select coterie of friends and our ways of life, is not ecologically sound. If we are to have a quality future, all mankind must be in on its creation.

What are the obstacles to synergy in our contemporary industrial world? In recent history we find that the greatest threat to context, and thus synergy, is one-sided adherence to the competitive model of industry in which most often if one wins the other loses.

These zero-sum games are not conducive to building the high synergy society described by Ruth Benedict. One is, as it were, precluded from cooperation in advance, except in the form of monopoly which is harmful to the public. Additionally, competition had made us rely on complex industrial technology, forcing us into a mold of life in which the need for technical order makes unidimensional what was to be a pluralistic society. There are countless examples of negative consequences for behavior because of the man-made environment, due to misapplication of science and technology (35, 36). How can the individual contribute to high energy, minimally labor intensive, methods of production? Our present societal course increases deadening man/machine interactions and decreases the joy of synergy.

The third category of obstacles to synergy is psychological. In our early search for biological security we accomplished a freedom from external conditions with designs for shelter from the elements, to guarantee food supply, etc. But these designs brought about a widening gulf between public and private life that can only be overcome by accomplishing freedom from internal conditions, an inner liberation. Understanding our selves has always been difficult when preoccupied with externals. Presently, the stumbling block to a complementary inner liberation is our reliance on external technology and wealth. Following Masserman, I hypothesize that this is a consequence of our manipulation of the environment by "Ur-delusions" (23, 45). These are ways of perceiving and conceiving basic defense mechanisms against intrapsychic fear on each of the three brain levels, arising from the realization of the frailty of our selves. In order not to be overcome by existential anxiety, we must have early on formulated conscious rationalizations which have been transformed into the basic assumptions of every growing child.

1. On the first brain level, the Delusion of Immortality is needed to deal with the realization that everyone dies and that survival is of the species, not of the individual. To enable planning of day-to-day activities, each of us will have to assume that we will not die, not now at least.

2. On the second brain level, the Delusion of Universal Friendliness, with its underlying inevitable fear of others, exists to overcome the realization that there are strangers outside of our primary group. If we want to develop communalities, we must assume that we are not working with enemies, at least not as long as they are talking to us.

3. On the third brain level, the Delusion of a Deus ex Machina, or Omnipotent Servant, is to overcome the realization of our limitations. Without the belief that somehow eventually someone will solve any and all problems, we would endanger our creative

capacity. This delusion has been successful in the form of religion or a system of government or any other societal structure. Today especially alive in most minds is the delusion that science and technology will take care of the world's problems and that salvation lies in (industrial) growth.

The three Ur-delusions interlock as do the three brains. But the apparent success of the industrial paradigm as the servant of the third Ur-delusion that gave the West its power makes change practically impossible, especially in parts of the underdeveloped world. We are only beginning to realize that there are limits to what science and technology can do: there are limits to growth. We are rediscovering that the principal aim of the intellectual brain is to assist in the humanization of mankind, not to provide material wealth. Man does not live by bread alone, and although money can buy privacy, it cannot buy community.

The awareness that we have to change the predominant third Ur-delusion, as the paradigm whereby we live, has been focused by two recent developments. The first is a feeling of relativism, after devastating world wars shook our emotional commitment to technical progress and the discovery of the relativity principle in physics (traditionally regarded the "hardest" of sciences) brought our materialistic approach into question. The second is the global need to discover radically new communalities with other than the traditional business or political goals, exemplified by the minorities' search for liberation. Previous feelings of identity with family, friends, entrepreneurial or institutional contacts are now superseded by bonds on the basis of age (40), sexual orientation, race or ethnicity (rather than tribe or country), or perceived roles (singles, mental patients). There is increasing discontent with traditional alliances and a search for conceptual and interaction spaces in which to "be their own." We might call this a search for conscious (or designed) communal identity, which reinforces the feeling of relativism. With a knowledge of the relative value of any single approach to life and an overriding urge for new communalities, people again begin to express their wanting to be recognized individually as being useful for the whole of mankind. In my opinion, continued evolution will be possible because of our understanding that giving of ourselves contributes to the essence of mankind. Lewis Thomas expressed this by remarking that language is the ecological niche created by the human species. We show our yearning to be socially useful by wanting to talk to each other (58). In the terminology of this chapter: we are useful by potentiating each other's conceptual space. This explains the contemporary interest in innovative designs for communality, e.g., the human potential movement, experiential growth workshops and mass festivals like Woodstock. People are interested in becoming and then sharing the evoked, heretofore unexperienced, parts of themselves.

But this means a change of priorities, a focus on moral issues when we determine new communalities. According to Redfield, technical order was subordinated to moral order in original communities (55). And in our society too, knowledge should again be of what is worth doing. As Medawar said, ". . . we should get used to the idea that moral judgements should intrude into the execution and application of science at every level" (49, p. 501). Planning for a post-industrial, multicultural society needs a radical restructuring of values, and Churchman recently addressed the need for a design methodology in which requirements in traditional planning are complemented with individual and social morality (11). Harman talks of the need for a "New Age" paradigm, including (1) a metaphysic asserting transcendent man, and (2) a person-oriented society. He feels that such a paradigm runs fundamentally counter to our present industrial thinking and that we, therefore, must learn to choose between policies based on extension of our present world and those needed for a new age. He cautions:

> Thus at the least, it would seem prudent to test policy decisions both against the eventuality that the view presented here may prove accurate, and also against the opposite eventuality, that it may simply turn out to be wrong and our current travails will be interpreted in some other way.
>
> Under the assumption that the paradigm-shift interpretation is more or less correct (that is, that the shift seems possible and desirable, but by no means automatic), it follows that the main challenge to society is to bring about the transition without shaking itself apart in the process. Every major policy decision tends either to foster the change or to impede it. Actions which attempt to force it too fast can be socially disruptive; actions which attempt to hold it back can make the transition more difficult and perhaps bloody. For example, there can be little doubt that maintenance of strong economic and legal-enforcement systems through the transition period is essential; yet these systems too must be flexible to change. Seldom in history has such delicacy of balance been required, to achieve a major social transformation rapidly and yet not rupture the social fabric.
>
> (33, p. 217)

We need to realize that more important than intellectual comprehension is the emotional acceptance of a transformation into

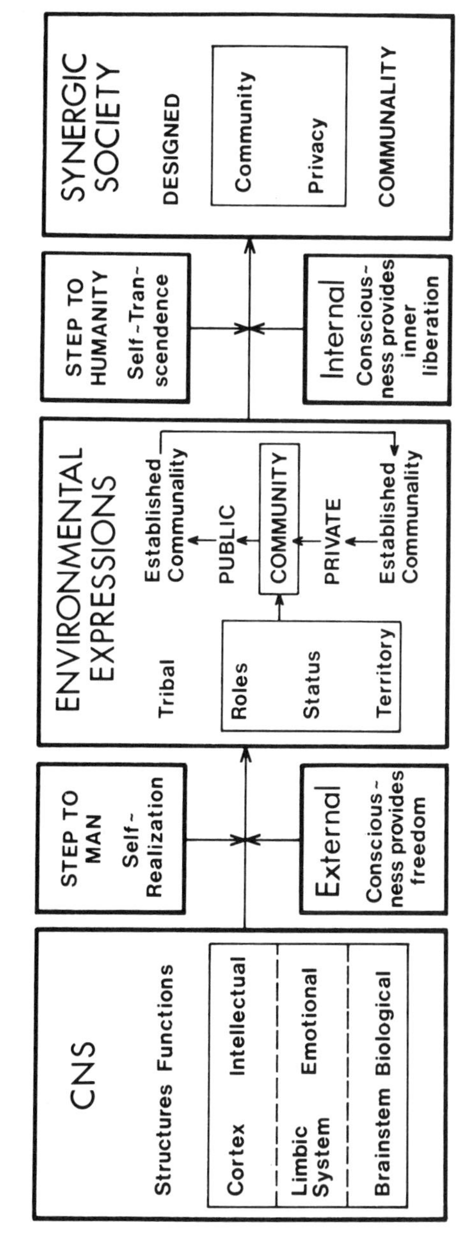

Figure 4:
Privacy and Community after Transcendence of Self

becoming humane. To accomplish this we need to provide building blocks or precursors for imaginative change. The model of humane evolution proposed here, as well as the framework in which suggestions for a synergic design process are presented, should be judged by the emotional reactions they evoke. Study of emotional-cognitive structuring has shown the intimate connection between the quality of the emotion evoked by an idea and the precision of its memory (30). And on the basis of the brain/behavior/environment theory proposed here, it may be reasonably expected that collectively conceptual changes are brought about in similar fashion. Once people "buy the idea," there are sufficient specific proposals for synergic post-industrial non-zero sum games to change economics and policy quickly (28, 33, 56).

In sum, it means that we should once again fit into the scheme of this universe, rather than view the environment as something to be conquered and molded into our image. In this sense, the implications of re-tribalization, as Goldsmith calls it (28), are worth consideration. Additionally, we should allow ourselves to grow, to explore our humane potential, and to design environments in which we may be replaced by the commitment to help *each other* and thereby to develop one's self. As E. Becker expressed it, we must understand and practice the art of giving again (6, p. 27 ff). A gift is the basis of our own life, and thus is the act of giving the basis of community.

The flowchart in Figure 4 shows how our becoming *human* fractured the environmental expressions of private and public life, and that our becoming *humane* leads to the design of contexts for simultaneous community and privacy. The step to man implied risking previously existing synergy in order to learn to use cortical tools consciously and conquer the environment. But the consequence of behaving differently than before in the environment is diminished context. Freedom of environmental constraints then allowed initiation of divergent ways of reconstructing context, and fragmentation of behavior and place increased. Private and public domains became differentiated in a vicious cycle of increasing specialization. This inner split can only be reconciled if increased consciousness of our own needs and feelings enables transcendence when needed for a mutually beneficial cooperation with others. When we talk about transcendence of self, we are referring to the complement to realization of self, and we seek inner liberation as a complement to freedom from environmental constraints. In either case, a complementary way of being emerges: altruism to make up for egotism, cooperation to balance competition, wholistic as well as reductionistic application of logic. When we become humane, we begin to design for each other what we would design for ourselves. We don't yet know how our synergic society will look like, but we can articulate some principles to begin the process of its design.

TOWARD A FRAMEWORK FOR SYNERGIC DESIGN

Human evolution proceeds by challenges and responses. This is the reason why the Chinese iconograph for danger means opportunity as well. Synergic designs have been reactions transcending individual or disciplinary interests. Challenged by the outcome to the Sputnik, we landed on the moon because of synergy between our competing Armed Forces, forced to cooperate in the design of NASA. David, the first baby growing up in the totally artificial environment of a sterile bubble, has as of now, at age six, belied the predictions that he would be emotionally disturbed because of lack of direct contact in his early years. His survival and thriving up till now are the result of a synergy between many scientific and technical disciplines and myriad moments of tender loving care on the part of people whose behaviors must have transmitted meaning, notwithstanding being out-of-place. There was designed communality and it apparently provided sufficient context.

Both examples show how much conscious cooperation and empathic understanding is required from specialists with divergent backgrounds. They all brought to the task an image of what had to be accomplished, and that kept the synergy going. It is true that once the goal has been achieved, relaxation and complacency sets in and synergy disappears. How are we to treasure the image of a process, in addition to a product of *partnership*, so that the pursuit of the common good may be continuous?

Obviously, our approach must be systematic. But it cannot just be the systems approach to the functioning of an ongoing system, since it has to deal with process. Thus, it must address *system formation*, in which there is constantly breakdown and new construction of elements, as Gray described the process of emotional-cognitive structuring necessary for knowledge growth (30). Anyone who has intimately lived with another knows how difficult is the essential daily renewal of relationships. One needs imaginative tolerance, suspension of disbelief (as in the theatre) and most of all, empathy. I am indebted to Koch and Xhignesse for assistance in developing a Gestalt (39) which simultaneously depicts individual brain processes, psychological attitudes and societal structures, thereby compressing and enriching the information of Tables 1 and 2. Figure 5 superimposes Jung's classification of psychological types related to function in society on a chart of left and right brain hemisphere differentiation. The functions are: (1) (biological) enterprise and (social) governance in the right hemisphere, reflected in the sensate and feeling types, and (2) (intellectual[13]) artistic

---

[13]To accomplish symmetry and to do justice to obvious Western cultural differentiation, the intellectual pursuit of the Arts has been separated from that of the Sciences and Technologies.

Figure 5:
INDIVIDUAL TYPOLOGY, SOCIETAL STRUCTURES AND CEREBRAL HEMISPHERES

LEFT (right side of the body)	RIGHT (left side of the body)
Conceptual	Biosocial
VERBAL	NONVERBAL
Sequential	Simultaneous
TIME	SPACE
Analytic	Synthetic, Gestalt
Linear, causal	Holistic, a-causal
Reading, writing, naming	Pattern recognition

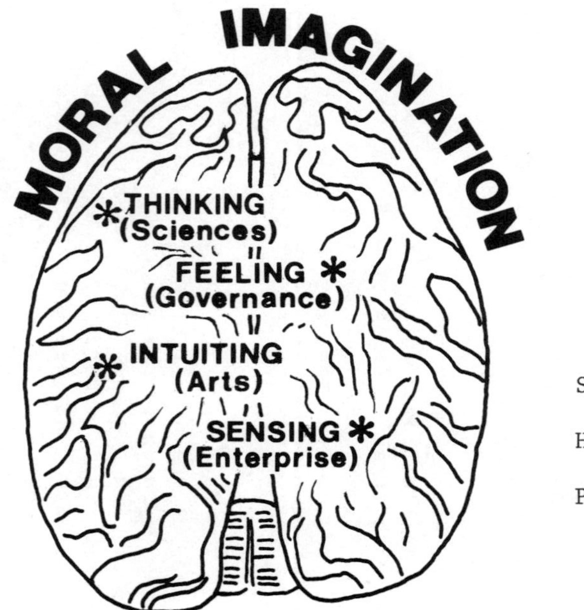

*The four major mental capacities according to Jung

and scientific-technological capabilities in the left hemisphere, reflected by the intuitive and thinking types. On the basis of our discussion of the human paradigm change, moral imagination, i.e., conceiving of what is worth doing, should guide the synergy between the functions of these structures. This type of imagination is placed in the frontal lobe with its potential empathy for images of value to others. Moral imagination should be kindled by curiosity about what unexpected results will come from our transcendence.

The two dimensions of Figure 5 do not allow depiction of perspectives resulting from a change of the viewing angle. But, if one thinks of the four main points in the hemispheres as corners of a tetrahedron turning around its axis while progressing in time, it becomes clear that the location of left and right is relative. Also, it becomes possible to describe synergy in terms of changing the mix of natural abilities (Jung's typology) or societal functions, thereby influencing the output in a non-zero sum game. For instance, if a person is predominantly a thinker paying very little attention to feeling, increasing his feeling life should not diminish his thinking ability but improve it, as shown in instances when intellectuals have become good politicians, the philosopher-kings of Plato. It is, after all, a fact that man is not only the most intellectual, but also the most emotional animal, which is implied in emotional-cognitive structuring theory.

The main point of Figure 5 is the idea of complementarity. If science and technology foster communality, art will promote community. If enterprise is the supreme expression of private interests, governance will represent the realm of the public. Complementarity applies to all cultural expressions. If ". . . the principal aim of culture and its raison d'etre is the humanization of man" (61, p. 3), the interdependence of cultures may become as essential to our survival as the interdependence of a world economy which made 18th century notions of the marketplace obsolete. We, therefore, must strive for a transformational society in which cultural pluralism is the moral order, needing a different type of planning and design than practiced today (11, 36, 61). I used the term, "designed communality," for one of the aspects of this needed direction, and I believe such planning and design to be realizable only if it is synergic.

Synergic design shares with many issues today the observation that it is easier to *pro*scribe than to *pre*scribe its features. Designers know, for instance, that one should *not* leave people out of the planning process, make unilateral decisions, limit information flow. Examples of non-synergic (or even negative synergic) design solutions with unwanted and unforeseen consequences proliferate. There is the demolition of the infamous Pruitt-Igoe public housing complex, the once-proud "community" project in the St. Louis of the fifties. The lessons of Pruitt-Igoe appear to have been ignored, e.g., the HUD New Towns, public housing of the seventies, are also stalled in their development, and there are the increasing costs of guarding and patrolling shopping malls and apartment complexes. In suburbia we witness the preference for privacy (territorial feelings are enhanced by plots for individual homes) over community.

One thing is certain from the above. Awareness of the deepest sources of knowledge, fears and assumptions of man is needed if one wants to determine features of synergic design. The designer can then fit the scientific and technical knowledge into a framework of meaning (1) for the individual in the function of CNS levels and their Ur-delusions and (2) for the society in terms of the prosthetic brain counterparts. The designer can be alerted to the fact that growing up in a culture, following the rules of a community, provides a context in which design decisions are made on the basis of hidden assumptions, unquestioned (and often unconscious) ways of doing things (13). Table 3 summarizes the conceptual steps in attending to the Ur-delusions. The designer should transcend the universal longing for immortality with the provision of positive environmental features intended to optimize health and welfare. Entrepreneurial considerations should be balanced by consideration of biological, emotional and social impacts. The designer should replace the delusion of friendship with, or being pleasing to, everyone by an awareness of where one's design fits in the overall scheme of special interests and privileges. In governance terms, the designer should be constantly in tune with the political scene, lest plans fall prey to partisan expediency. Finally, to keep abreast of what is going on in conceptual space, in addition to employing superconsultants, the designer simply needs to keep an open mind.

In practice, I may point to four areas of consideration in synergic design: evocative participation, appropriateness, multicultural standards, and a multidisciplinary systems approach. Some of these practical insights were arrived at by the builders of Utopia (34), others derive from a general systems approach.

The *participatory process* entails learning to like self-organization. Firstly, nobody can work synergically unless first-hand knowledge of the meaning of authority has been obtained. I would call this learning to master your own limbic system before you attempt to guide others, which might be learned by experiencing group relations systematically.[14] Thereafter, you can easily

---

[14]For instance, with the A.K. Rice Institute, whose *Group Relations Reader* (1975) also contains articles on design and planning.

TABLE 3

CONSCIOUS KNOWLEDGE AND SYNERGY

BRAINS	KNOWLEDGE OF	SOCIETAL STRUCTURES	UR-DELUSIONS	SYNERGIC DESIGN RESPONSE
Prosthetic	Limits	Arts, Sciences, Technologies	Omnipotent Servant	Keeping an open mind
Social	The Strangers	Governance	Brotherhood of Man	Staying attuned to the political situation
Biological	Physical Death	Enterprise	Immortality	Promotion of health and welfare

understand and may want to apply methods and procedures proposed for synergic group functioning (e.g., 12, 14). Secondly, learning to fit oneself and grow into an organization can occur when an individual works at and perfects a part of the design, while adhering to the communal effort. An example would be user participation in design via scale modelling (18). This leads to the central point in self-organizing systems: efforts at self-help can be evocative of heretofore hidden aspects, i.e., they lead to individual growth. Thus far in Western society, the preponderant direction of design for communal purposes has been to facilitate production. This has led to such more and more specialized environments as factories, or hospitals, or retirement villages. But if we were to design communality from a synergic perspective in which all parties and each factor is considered, we would come up with contexts not solely prescriptive of function, but also evocative and supportive for a range of behaviors and moods. Izumi illustrates this difference by contrasting the restrictive, solely reverential purpose of a Gothic cathedral with the infinite variety of experience possible in a Japanese teahouse (35, discussed in 24). Additional examples of design for communality with a view toward allowing growth of diversity are Paolo Soleri's Arcosancti and the heterogenistic colony of Maruyama (46). Important is the earlier mentioned tactic of keeping the design open-ended as long as possible, thereby allowing reflection of ongoing collective processes. This also prevents *a priori* classification, which serves reductionism and thereby a design that possibly denies the use for which it was intended (examples in 24, p. 21, ff).

Appropriateness in design can be brought about by simplification and search for appropriate size. These practices will increase possibilities for user participation and economy. Each user craves uniqueness, but we do not need complex design, since it can be achieved equally well by standardization (24, 35). Another element of simplification in design is proper environmental coding to help establish context (18, DeLong this volume). It is part of an empathic approach to design for man-environment relations, which also includes prosthetic, simulative and futuristic aspects (25). As to the choice of size for a design, examples of counterproductive effects of large-scale technology and oversized human settlements abound in the literature.

Setting *multicultural standards* is the most complex issue in synergic design. It encompasses the notions of (a) turning the abstract into concrete guidelines, (b) complementarity, and (c) networking.

(a) In writing standards, one tries to concretize general insights into specific directions without opening oneself up for abuse by designers of specialized or exclusive environments. Let us take the standard of physical proximity to establish community. One could take the strengthening of face-to-face interaction as the cue, as did Alexander in proposing short travel times to work, local schools and food production (to encourage ecological fit), etc. (1). On the other hand, there is the need to counteract purely local relationships with the participation into greater, even global communalities. Erikson in his recent Jefferson Lectures for the National Endowment of the Arts, refers to Jefferson's ambitions as an architect and planner: ". . . which reminds me once more of Jefferson's idea of the ideal size of a township. He may not have been so far off in regard to the optimal communal units to be built into a future megalopolis (except that bicycle paths rather than footwalks may there be the healthy measure of optimal surroundings). The main point is that at each stage of life there must be a network of direct personal and communal communication safely set within the wider networks of automobility and mass media." (19, p. 123) What we, therefore, must search for are counterparts in the environment to what people perceive as needs for communal identity. As we discussed earlier, such environmental counterparts cannot be found, for instance, in strengthening the ethnicity of neighborhoods or providing for adolescent hangouts in our shopping centers, even if these design measures are necessary. Thus, we see the need for the next aspect of multicultural standards: complementarity.

(b) It was already noted in Utopias that community and privacy both had to be provided if the settlement were to function. One design implication was that communal territory should not be erected by taking away private territory (34). I would stress the transformation aspects of privacy on different CNS levels (26) and count on education as part of any design implementation. But the need for complementarity goes way beyond community and privacy. In general, one needs homogeneity *and* heterogeneity in expanding conceptual space while maintaining biosocial security and familiarity. Design should incorporate different hierarchies and stages of development; for instance, old-age villages and one-company towns are not conducive to synergy. In other words, this again goes back to not planning special purpose environments. This principle is in contrast with traditional Western logic, which tries to achieve unity through homogeneity (46, Izumi in 24). This brings up the earlier mentioned aspect of the new paradigm of man: responsibility for creating a new society while using all parts of the CNS, in particular, the synergy of the spatially oriented wholistic right hemisphere with the time-conscious detailed analytic reasoning of the left hemisphere (see Figure 5). Attempts at literally "putting it all together" have already been introduced, e.g., standards for environmental assessment in the Department of Housing and Urban Development refer to what we now may call synergic design concepts, such as "sense of community" and "psychological well-being" (e.g., 41).

(c) Finally, one needs to pay attention to how information on the many guidelines for different cultural settings and the complementarity of divergent insight will be made available. I believe that one of the attributes of the communications era we have entered is the development of information networks. Presently, these are mostly informal ways of like-minded people to stay in touch, but we are beginning to see the use of networks as generators and purveyors of standards. Some of the resources mentioned under the next area to be discussed expressly state that multidisciplinary issues are resolvable only by networking (e.g., 5). This applies specifically to design problems heretofore tackled only with information coming from science and technology. The resulting design solutions may need modification with reference to the infinite variety of personal experiences which are rarely addressed in formal communications (24, p. 119).

Our creations are studied today by the *general systems approach*, since their complexity overwhelms the capability of understanding by any one individual. For synergic functioning, it is equally important to take into account that there are "systems of nature" as well as the unspoken, informal culture, which go beyond our verbal reasoning ability (32). In design based on brain/behavior/environment relations these systems are part of our dracolimbic context (31). Involvement of nonrational as well as rational human systems makes a multidisciplinary approach mandatory without *a priori* exclusion of any insight. The empathic design approach will mean that much of our future environmental organization goes hand in hand with economic and social reorganization. If we are to program our designs in this comprehensive fashion, we need a taxonomy in the area of relationships between cognition, behavior and environment. Proposals have been made by the military (e.g., 16, 17) and individual members of the Environmental Design Research Association (53), the Association for the Study of Man-Environment Relations (5) and the Design Methods Group (29). As said earlier, all such proposals need to be viewed in the light of the anticipated changes in our value system (11, 28) and our moral imagination. We need to cast about for new methods, in different disciplines and in different cultures. A cross-disciplinary example is the attention on the concept of altruism, renewed by animal studies in ethology and social biology. For transcultural examples, I refer to the Indonesian practices of *musjawara* (implementing decisions only if there is total consent) and *gotong rojong* (collective self-help, similar to the old neighborly custom of barn-raising).

We have discussed some presently recognized aspects of synergic design, but we should know that the synergic process can work only if it becomes a way of being conscious, to which society becomes receptive. This framework will further develop our understanding and structuring of continually evolving concepts: family, individual, privacy, community, identity and others so basic to the endeavor of this chapter. Meanwhile, we know that if empathy and synergy are not welcome today because society is not ready for designed communality, their future lies in our subjective commitment.

REFERENCES

1. Alexander, C. The city as a mechanism for sustaining human contact. In W. R. Ewald, Jr. (Ed.), *Environment for Man*. Bloomfield: University of Indiana Press, 1967.

2. Altman, I. *The Environment and Social Behavior*. Monterey, Cal.: Brooks, Cole, 1975.

3. Ardrey, R. *The Social Contract*. New York: Atheneum, 1970.

4. Aries, P. *Centuries of Childhood: A Social History of Family Life*. New York: Vintage Books, 1962.

5. Beck, R. J. User generated research in function requirements for roles engaged in the design, management, building, teaching, informing and use of the environment. *Man-Environment Systems*, 1976, *6:* 151-162.

6. Becker, E. *Escape from Evil*. Glencoe, Ill.: The Free Press, 1975.

7. Bossley, M. I. Privacy and crowding: A multidisciplinary analysis. *Man-Environment Systems*, 1976, *6:* 8-19.

8. Brenner, M. H. Estimating the social costs of national economic policy. Paper #5, Joint Economic Committee, Congress of the U.S., October 26, 1976.

9. Calhoun, J. B. Environmental design research and monitoring from an evolutionary perspective. *Man-Environment Systems*, 1975, *4:* 3-30.

10. Chermayeff, S. and Tzonis, A. *Shape of Community*. Baltimore, Md.: Penguin Books, 1971.

11. Churchman, C. W. Morality and planning. *Design Methods and Theories*, 1976, *10:* 165-181.

12. Coulter, N. A. *Synergetics: An Adventure in Human Development*. Englewood Cliffs, N. J.: Prentice Hall, 1976.

13. Colman, A. D. Irrational aspects of design. *Man-Environment Systems*, 1973, *3:* 161-176.

14. Craig, J. H. and M. *Synergic Power: Beyond Domination and Permissiveness*. Berkeley, Cal.: Pro-active Press, 1974.

15. Davis, G. and Altman, I. Territories at the workplace: Theory into design guidelines. *Man-Environment Systems*, 1976, *6:* 46-53.

16. Davis, T. A. Evaluating for environmental measures. In J. Archea and C. Eastman (Eds.), *EDRA Two*. Stroudsburg, Pa.: Dowden Hutchinson and Ross, 1970.

17. Davis, T. A. Systematizing man-environment information. *Man-Environment Systems*, 1974, *4:* 181-184.

18. DeLong, A. J. Coding the environment. In W. F. E. Preiser (Ed.), *Psyche and Design*. Orangeburg, N. Y.: ASMER, Inc., 1976.

19. Erikson, E. *Dimensions of a New Identity*. New York: Newton, 1974.

20. Esser, A. H. Social pollution. *Social Education*, 1971, *35:* 10-18.

21. Esser, A. H. Evolving neurologic substrates of essentic forms. *General Systems*, 1972, *17:* 33-41.

22. Esser, A. H. Structures for man-environment relations. In W. F. E. Preiser (Ed.), *Proceedings of the Environmental Design Research Association*, 4th Annual Meeting. Stroudsburg, Pa.: Dowden Hutchinson and Ross, Inc., 1973.

23. Esser, A. H. Social pollution in the evolution of mankind's communal imagery. In J. White (Ed.), *Frontiers of Consciousness*. New York: Julian Press, 1974.

24. Esser, A. H. (Ed.). *Health and Built Environment*. Ottawa, Canada: Health and Welfare, Canada, 1974.

25. Esser, A. H. Design for man-environment relations. In A. Rapoport (Ed.), *The Mutual Interaction of People and their Built Environment*. The Hague: Mouton, 1976.

26. Esser, A. H. Theoretical and empirical issues with regard to privacy, territoriality, personal space and crowding. *Environment and Behavior*, 1976, *8:* 117-124.

27. Fuller, B. *Synergetics: Explorations in the Geometry of Thinking*. New York: Macmillan, 1975.

28. Goldsmith, E. *The Stable Society*. Wadebridge, U.K.: Ecosystems, Ltd. (in press).

29. Grant, D. P. *Systematic Methods in Environmental Design: An Introductory Bibliography*. Monticello, Ill.: Council of Planning Librarians, 1972.

30. Gray, W. Emotional cognitive structures: A general systems theory of personality. *General Systems*, 1973, *18:* 167-173.

31. Greenbie, B. B. *Design for Diversity*. New York and Amsterdam: Elsevier, 1976.

32. Hall, E. T. *Beyond Culture*. Garden City, N. Y.: Doubleday, 1976.

33. Harman, W. W. Planning amid forces for institutional change. *Man-Environment Systems*, 1972, *2:* 207-220.

34. Hayden, D. *Seven American Utopias*. Cambridge, Mass.: The MIT Press, 1976.

35. Izumi, K. The (in)human(e) environment. Transcript of six television lectures. Ottawa: Ministry of State for Urban Affairs, 1973.

36. Izumi, K. Cultural issues and man-made environment. Position paper. Canadian Commission for Unesco, Calgary, Alberta, Canada, 1977.

37. Jung, G. C. *Two Essays on Analytical Psychology*. Cleveland: Meridian Books, 1965.

38. Kaplan, A. Maslow interpreted for the work environment. *Man-Environment Systems*, 1976, *6:* 246-248.

39. Koch, T. and Xhignesse, L. Personal communication. Stanley House Conference on Design of Multicultural Environments, New Richmond, Quebec, 1977.

40. Latowsky, E. and Kelner, M. Youth, the new tribal group. In A. Davis and P. Herman (Eds.), *Social Space: Canadian Perspective*. Don Mills, Ontario: Free Press, 1969.

41. Lemer, C. A. Environmental impact, environmental control and the performance concept: Some thoughts on putting it all together. *Man-Environment Systems*, 1976, *6:* 86-90.

42. MacLean, P. D. The brain in relation to empathy and medical education. *Journal of Nervous and Mental Disease*, 1967, *144:* 374-382.

43. MacLean, P. D. On the evolution of three mentalities. *Man-Environment Systems*, 1975, *5:* 213-224.

44. Maslow, A. H. Synergy in the society and the individual. *Journal of Industrial Psychology*, 1964, *20:* 153-164.

45. Masserman, J. H. *Principles of Dynamic Psychology*. Philadelphia, Pa.: Saunders, 1946.

46. Maruyama, M. Cultural factors in high density habitats. In W. F. E. Preiser (Ed.), *Psyche and Design*. Orangeburg, N. Y.: ASMER, Inc., 1976.

47. McBride, G. Privacy: A relationship model. *Man-Environment Systems*, 1977, *7:* 145-154.

48. McIver, R. M. *Society, Its Structure and Changes*. New York: Smith, 1931.

49. Medawar, P. Does ethology throw any light on human behavior? In P. P. G. Bateson and R. A. Hinde (Eds.), *Growing Points in Ethology*. New York: Cambridge University Press, 1976.

50. Nisbet, R. A. *Community and Power*. New York: Oxford University Press, 1962.

51. O'Neill, J. Public and private space. In A. Davis and P. Herman (Eds.), *Social Space: Canadian Perspective*. Don Mills, Ontario: Free Press, 1969.

52. Porteous, J. D. *Environment and Behavior*. Reading, Mass.: Addison-Wesley, 1977.

53. Preiser, W. F. E. (Ed.). *Programming for User Needs.* Stroudsburg, Pa.: Dowden Hutchinson and Ross (in press).

54. Rapoport, A. *Human Aspects of Urban Form.* New York: Pergamon Press, 1977.

55. Redfield, R. *The Primitive World and Its Transformations.* Ithaca, N. Y.: Cornell University Press, 1968.

56. Schumacher, E. F. *Small is Beautiful: A Study of Economics As If People Mattered.* New York: Harper and Row, 1974.

57. Suttles, G. G. *The Social Construction of Communities.* Chicago: University of Chicago Press, 1972.

58. Thomas, L. *The Lives of a Cell.* New York: The Viking Press, 1974.

59. Tibbetts, P. and Esser, A. H. Transactional structures for man-environment relations. *Man-Environment Systems,* 1973, *3:* 441-488.

60. Westin, A. F. *Privacy and Freedom.* New York: Atheneum, 1967.

61. Wojciechowski, J. The nature of cultural pluralism and national identity. Position paper. Canadian Commission for Unesco, Calgary, Alberta, Canada, 1977.

PRIVACY, COMMUNITY AND ACTIVITY SPACE: A RELATIONAL EXPLORATION[1]

Allen A. Turnbull, Jr.

Carleton University
Department of Psychology
Ottawa, Canada K1S 5B6

ABSTRACT

"Privacy," "community" and "activity space" are frequent topics of discussion among social scientists, philosophers, lawyers, architects and urban planners. The task of coming to an unambiguous and comprehensive understanding of either privacy (19), community (7) or activity space (5) presents a formidable problem. Understanding the complexities of the interrelationships among privacy, community and activity space presents even more difficulties (6, 15, 17, 30, 31).

The first part of this paper presents an overview of privacy, community and activity space as individual concepts. Working definitions are proposed, followed by a discussion of the possible interrelationships among the three concepts. The second part presents the results from an ongoing research project that is investigating these interrelationships through the development of an attitudinal inventory (32). This modest beginning will, it is hoped, lead to elaboration of a more comprehensive model of the privacy-community-activity space interrelationships in the future.

[1]This research was supported in part by Canada Council Grant S74-0371. Appreciation is expressed to Lloyd Strickland, John Barefoot, Bryan Laver, Clarence Aasen and Richard Davies for their advice and assistance during various phases of this research project.

WORKING DEFINITIONS AND POSSIBLE INTERRELATIONSHIPS

Privacy

Altman isolated several common themes that seemed to emerge from previous discussions of privacy (1). The primary themes were: 1) the ability to control 2) input and output through 3) the variation of interpersonal boundaries. Most definitions of privacy center around the first theme, control and freedom of choice, by including such phrases as "the individual's right to control" (3), "freedom of the individual to pick and choose" (29), "ability to control interaction" (28) and "obtaining freedom of choice" (27).

The second theme is best categorized as control over *output* to other persons and control over *input* from other persons and the physical environment. Some general inputs and outputs have been normatively defined by a particular culture as relating to privacy. Kelvin has emphasized the importance of such cultural norms in determining what will be labeled as "private" (16). Moreover, within a particular culture, personal idiosyncracies will also influence categorization of particular inputs and outputs as private; what might be considered private or an invasion of privacy by one person may not be so classified by another. The phenomenology of the individual must be considered in any analysis of privacy.

The third aspect of privacy is the *mechanism* through which a person controls the "whats" he has labeled as private. Altman discussed the idea of an interpersonal boundary control process by using the analogy of a cell membrane, which can be differentially permeable depending on the level of privacy desired (1). The permeability of an individual's interpersonal boundary is constantly changing, regulating the frequency and intensity of the various inputs and outputs. A variety of behaviors--verbal, nonverbal and environmental--aid in determining the momentary permeability of one's interpersonal boundary.

For the purposes of this investigation, Altman's definition of privacy as "selective control of access to the self or to one's group" (1, p. 18) was used, although individual privacy rather than group privacy was the primary focus of analysis. This definition was broad enough to include the various types of privacy proposed by Marshall, e.g., seclusion, solitude, anonymity, self-disclosure, intimacy and not neighboring (20).

Community

Prominent themes in the vast literature on community revolve around 1) community as a geographical locale, 2) community formed through the integrative behavior of primary and secondary groups and 3) community as the source of a psychological sense of wholeness and attachment for an individual (7, 12, 17). Dentler thus defined community as "a place within which one finds all or most of the economic, political, religious and familial institutions around which people group to cooperate, compete or conflict" (9, p. 16). Many similar definitions include each of the three themes; however, another way of viewing community is through a word which shares its common root--communion.

Minar and Greer stated that community "expresses our vague yearning for a commonality of desire, a communion with those around us" (25, p. ix). Keyes suggested that communion with those around us results from being "known whole" (17). A person is known whole through involvement with many different individuals who all know each other. Involvement in such a closely linked network of friendships should foster high levels of openness and honesty. Keyes argued that modern man is only partly known by most of his associates and consequently suffers from loneliness and rootlessness. For him, it is essential for the development of "positive mental health" (13, 22) that persons be known whole by some person or group.

Slater expressed similar sentiments in a discussion of open networks; that is, networks in which the individuals involved only know the central person and rarely know each other. The result of such a friendship pattern is that "everyone controls her own social milieu . . . and she is never forced to integrate the disparate sides of herself but can compartmentalize them in disconnected relationships" (30).

The theme of wholeness was also discussed by Poplin:

> Community seems to involve a sense of identity and unity with one's group and a feeling of involvement and wholeness on the part of the individual. In short, the term *community* has been used to refer to a condition in which human beings find themselves enmeshed in a tightknit web of meaningful relationships with their fellow human beings. (26)

The working definition chosen for this investigation resembled Poplin's; it was "selective engagement in a close-knit web of meaningful relationships with others."

Note that no mention of physical space is made in this definition of community based upon the notion of communion. The inclusion of a geographical component in a theory of community is only an assumption. Clark echoed this criticism of most community definitions by writing that "to argue that place *influences* community is a very different matter from assuming that certain geographical units or areas are *synonymous* with it" (7, p. 398). Other community researchers (33, 34) have argued that the increased mobility and communication possibilities of the modern city dweller now allows for "community without propinquity." In any event, while it may be more likely for a close-knit web of meaningful relationships to evolve within a local context than within a larger urban context, this remains an empirical question and not a matter of definition.

Activity Space

The concept of activity space seems to be much simpler conceptually than the preceding concepts of privacy and community. The working definition for this study was "the geographical area within which one chooses to conduct most of his activities." Activities in this case refers to typical daily events such as working, attending school, worshipping, shopping, banking, eating, visiting with relatives and friends and general recreation.

The activity space within which we live--within which we can be known whole--has often been identified as our neighborhood. In the past it may have been the case that our sense of community grew out of our neighborhood; indeed, Keller defined neighborhoods as "local areas that have physical boundaries, social networks, concentrated use of area facilities and special emotional and symbolic connotations for their inhabitants" (14, pp. 156-157). If we accept this definition of neighborhoods, we may find that there are fewer neighborhoods today than there were a decade ago. Perhaps close-knit meaningful relationships were once found in the neighborhood, but as urban areas have evolved into larger open systems, and as individual mobility has increased, this may no longer be so. A nostalgic feeling for the typical neighborhood of yesteryear pervades much of the writing on community, but the modern technological city may allow community to be established without regard to geographical space. Recently, urban planners have incorporated the notion of local activity patterns into the design of new towns, a notable U. S. example being Columbia, Maryland. How these new towns adequately accommodate the activity space needs of the residents remains to be seen (4, 18).

Privacy and Community

Margaret Mead argued that both privacy and community were basic human needs, but she implied a trade-off between them; i.e., the typical anonymity of the large city may be seen as a strength allowing the attainment of a high level of privacy, but it can just as easily be seen as a weakness undermining community if carried to excess (23). The notion of an inherent conflict between privacy and community is a recurrent theme in much social science literature.

Slater has been one of the more articulate voices presenting privacy and community as antithetical needs.

> Who would not like to believe, for example, in the possibility of a society that would maximize personal autonomy and relatedness at the same time? But there is no way to guarantee that one man's need to be alone will never coincide with his neighbor's need to be with him. Every society tends to protect one need more than the other in a given situation. Accepting the fact that the issue must always be negotiated between people--that privacy and community are antithetical needs and cannot be simultaneously maximized--leads to full and complete despair, the despair of dissillusionment.
> (31, pp. 3-4)

Keys argued that contemporary Western society tended to protect privacy more than community, and we are consequently witnessing a loss of community.

> Even as we hate being unknown to each other, we crave anonymity and rather than take paths that might lead us back together, we pursue the very things that keep us cut off from each other. There are three things we cherish in particular--mobility, privacy, and convenience--which are the very sources of our lack of community.
> (17, p. 15)

Slater continues this line of reasoning by suggesting that our longing for privacy is simply a reaction to our loss of community and concomitant inability to control our environment.

> The longing itself is not a fundamental or driving human motive, but a reaction to crowding, complexity, and social dislocation. Those who live in stable preindustrial communities have far less privacy and far less desire for it than we do. They feel less manipulated and intruded upon only because they can predict and influence their daily social encounters with greater ease. The longing for privacy is generated by the drastic conditions that the longing for privacy produces.
> (30, pp. 125-126)

Other critics of the modern technological city have implicated man's loss of control over his physical and social environment as a central factor in the breakdown of interpersonal bonds and the promotion of both voluntary and involuntary privacy (2, 24). Bookchin mirrored Slater's concern over escalating privacy needs. He argued that the "individual who withdraws into himself and his private concerns, who fortifies himself with social neutrality and civic indifference, all the more delivers his privacy to the invasive forces from which he tries to escape" (2, p. 83). This argument holds that urban complexity leads to a perceived loss of control due to stimulus overload, which in turn leads to higher levels of privacy in an attempt to reduce stimulus inputs, which in turn leads to an even greater loss of control; at this point, the cycle begins anew.

The thesis of this paper is that privacy and community, as defined herein, are independent of each other. Slater is correct in noting that we cannot simultaneously be alone and also be with our neighbor (31), but it does not follow from this observation that privacy and community as general concepts represent antithetical needs. The factor of *time* is crucial in determining the interplay between privacy and community behaviors. For example, a person might prefer a private house secluded from other houses so that he might better control various inputs from and outputs to his neighbors. At the same time, however, this person may like to engage in a close-knit web of meaningful relationships by visiting his friends often and by having his friends visit him often. He may even tell his friends to drop by unannounced whenever they feel like visiting. Clearly, privacy and community need not be in conflict in this situation.

This example also illustrates the formation of a "we" feeling among one's close associates (8) which allows for an increased permeability within the close group of friends but not in relation to outsiders. Now, interactions within the group may be a part of each individual's "backstage" (10). As a consequence, community ensues with no large concomitant decrease in individual privacy. A phenomenological perspective would hypothesize that the perceiver would not perceive any conflict between privacy and community in this situation.

Privacy and Activity Space

Just as others have speculated that privacy and community are in conflict, Keller speculated that privacy and a local activity pattern are in conflict as well. She categorized neighborhood residents along a continuum from "reserved" to "sociable" and "local" to "urban":

> Each type has different, and perhaps incompatible, conceptions of privacy, space needs, and relations

> to immediate and more distant neighbors. An apt description of these types refers to the one type as "sociable," to the other as "reserved." The sociable resident moves into an area expecting to have warm, friendly relations with neighbors; the reserved type has no such desire. His need for privacy is stronger than his need for sociability, and the reserved resident primarily wants neighbors to respect this need. . . . Their spatial relations should be adapted to these deep seated preferences. The locally-oriented resident concentrates on the immediate local area for the satisfaction of basic needs--social, personal and material--whereas the urban-oriented type uses local facilities, services and contacts in a more limited and less exclusive way, essentially looking to the wider society for these things. The local resident resides in the city but lives in the neighborhood; the urban type resides in the neighborhood but lives in the city. How these two types relate to the differentiation between reserved and sociable is not yet clear, but the reserved type may be more urban, the sociable type more local.
>
> (14, p. 159-160)

It would seem that there need be no relationship between privacy and activity space since a particular desired level of privacy can be obtained through a number of behavioral mechanisms, only one of which would involve the spatial characteristics of one's activities. Hence, it was hypothesized that an individual's preference for a particular level of privacy would not be related systematically to his preference for a particular level of activity space.

Community and Activity Space

Until recently, few researchers have investigated the empirical relationship between community and activity space. As discussed earlier much speculation abounds but little data are available to clarify the issue. Wellman and his colleagues have begun to unravel the nature of the relationship (34). Wellman, Craven, Whitaker, Stevens, Shorter, DuToit and Bakker reported that only 13 percent of the average Toronto resident's intimate friends live in his neighborhood, only 25 percent live within the same borough, and about 25 percent do not live within the boundaries of Metropolitan Toronto (35). If community is defined as intimate social networks, Webber's "community without propinquity" (33) seems to be a reality.

As argued previously, their need to be no inherent relationship between community and activity space as theoretical concepts, so for this reason it was hypothesized that an individual's preference for a particular level of community would not be related systematically to his preference for a particular level of activity space.

ATTITUDINAL INVENTORY

Construction

The information concerning the relationship among privacy, community and activity space was obtained in the present study through an attitudinal inventory designed to measure individual preferences for various levels of privacy, community and activity space. The goal was to construct an inventory in which separate scales for privacy, community and activity space would be orthogonal to each other in Euclidian space. If this proved to be not possible, then this would call for a reassessment of the three independence hypotheses.

Two questions immediately arise. The first concerns individual preferences. Would individuals who prefer high levels of privacy typically prefer low levels of community; would individuals who prefer high levels of privacy typically prefer urban activity patterns; and would individuals who prefer high levels of community typically prefer local activity patterns? Specifically, what would be the nature of the correlations between the individual preference scales for privacy, community and activity space?

The second question concerns the actual behavioral relationship between these three concepts. An individual might *prefer* both a high level of privacy and community, but this combination might be impossible to *achieve* behaviorally. Consequently, two inventories were constructed. One measured desired levels of privacy, community and activity space (Desired PCA Inventory) and one measured achieved levels of privacy, community and activity space (Achieved PCA Inventory). A brief account of how these two inventories were constructed follows.

Item Sorting and Pilot Sorting

The first task was to collect a number of potentially useful items to be included on the PCA Inventories. These items were obtained from a number of sources. The literature on privacy, community and activity space was reviewed, and the many different aspects of these concepts were incorporated into possible items. Many of the original items from Marshall's Privacy Preference Scale (20) were included. In addition, many persons were asked to submit items that in their opinion would measure someone's preference for a certain degree of privacy, community or activity space. This procedure yielded 366 items that were included in the pilot sorting. Sorters were asked to sort the items into nine categories (see below) and comment on the ambiguity of the wording of the items. This pilot procedure resulted in the elimination of many poor items, so that the final sorting of items into nine categories was based on 239 items.

In order to choose items that were representative of only one of the three concepts, sixteen sorters read each item and placed it in one of nine categories, according to the three definitions given. The sorters were chosen to represent a cross-section of an urban population on the basis of age, sex, type of dwelling and place of residence. The definitions used for the item sorting were:

> *Definition A.* The person making this statement is indicating the degree to which he prefers to control (or prefers not to control) interaction, either verbal or physical, between himself and others.
>
> *Definition B.* The person making this statement is indicating the degree to which he prefers to engage in (or avoid) a close-knit web of meaningful relationships with others.
>
> *Definition C.* The person making this statement is indicating the degree to which he prefers (or does not prefer) to perform most of his activities within one small area (e.g., block or neighborhood) in which he resides.

Sorters classified each statement as best representing one of nine categories: (1) Definition A, (2) Definition B, (3) Definition C, (4) A and B, (5) A and C, (6) B and C, (7) All, (8) None, or (9) ?. The terms, "privacy," "community" and "activity space" were not used at any time in instructing the sorters.

A high degree of consensus existed for the classification of 92 items or approximately 30 for each definition. Items were retained for possible inclusion in the PCA Inventories if ten or more of the sorters indicated the same placement under Definition A, Definition B, or Definition C. Statements such as, "I would like to live in a commune," "I need places where I can go every day and be sure of seeing familiar faces" and "I would enjoy the anonymity found in a large city" were classified as relating to all three definitions. However, the large number of statements classified as primarily relating to privacy, community or activity space suggests that the central themes of these concepts might be orthogonal to each other. Obviously, conflict does occur among the concepts some of the time, but it is the thesis of this paper that the core dimensions are independent and that a conceptual model of the residential environment should take this into account.

Development of the Brief PCA Inventories

It was next necessary to choose the best 10-12 items for each of the three subscales to be included in the PCA Inventories. A limit of approximately 40 items was set to ensure that the PCA Inventory would be brief, easily administered and useful for field applications. In order to choose the items for the brief inventory, the 91 items classified as belonging to only one of the three definitions were included in a new inventory from which the final items would be selected. In addition, a social desirability measure was included to allow the elimination of any items highly correlated with it.

The subject sample for this phase consisted of 192 respondents from urban centers in Canada and the United States. A representative cross-sample was again obtained to ensure an equal number of males and females, apartment dwellers and equal representation from four different age categories ranging from 17 to 80 years of age. Respondents were asked to read each item and indicate the degree of their agreement by checking one of six possible alternatives ranging from Strongly Disagree to Strongly Agree.

After collecting the 192 completed inventories, the total sample was randomly divided into two subsamples, with some adjustment made to match for age, sex and dwelling classifications. A cross-validation technique identified the items having the desired psychometric qualities. The following criteria were used in selecting the final items:

1. An item must have a correlation coefficient of .20 or above on the appropriate scale;

2. An item must have a correlation coefficient of .20 or below on the two inappropriate scales;

3. The mean response for an item over all respondents must be between 2.0 and 5.0 on the six point continuum;

4. An item must not be correlated above .20 with the measure of social desirability; and

5. Over the entire scale, items should be representative of the full conceptual domain.

TABLE 1

Desired Privacy, Community and Activity Space Items

Privacy

1. I would like to have a private retreat which no one could enter without asking me.
2. I don't care if I can't be alone when I wish. (-)
3. When I go on vacations, I prefer to go to a secluded spot.
4. I want to have control over when I see other people.
5. I think that too much emphasis is given to the individual's right to privacy. (-)
6. To ensure privacy, I would want to live with a high hedge or wooden fence around my backyard.
7. I don't mind people knowing a lot about my activities. (-)
8. I am annoyed by people who try to get to know me better than I wish.
9. I like having people drop by my home without calling me beforehand. (-)
10. I would disclose my yearly income to anyone who wanted to know. (-)

Community

1. I would like to have a close group of friends with whom I do things regularly.
2. I would like to have a few close friends who all know each other well.
3. I would like being a member of a family that does things together often.
4. I could go a long time without socializing with other people. (-)
5. I need to be around people who accept me as I am.
6. I would like to be a member of a close-knit group.
7. I don't need to see my close friends every day. (-)
8. I can be the person I want to be without being a member of a group. (-)
9. I would rather work alone than as a member of a small work team. (-)
10. I like to engage in cooperative endeavors with others.

Activity Space

1. I don't care whether I live within walking distance of a shopping area. (-)
2. If I could do everything I wanted within my neighborhood, I would rarely leave it.
3. It is important that a grocery store be near my residence.
4. I would prefer to live where there is a movie-house nearby.
5. I dislike having to take the car to obtain the things I need.
6. I would enjoy frequent trips away from my neighborhood. (-)
7. I prefer to spend most of my leisure time outside my neighborhood. (-)
8. I would like to have a restaurant within walking distance of my home.
9. I want most of my friends to live outside my neighborhood. (-)
10. I want my job to be within walking distance of my home.

(-) indicates a reversed item.

This procedure yielded the 30 items listed in Table 1. The compound probabilities based on the item-total score correlations were less than .01 for all items except Item 10 on the Privacy Scale.

The items in Table 1 constituted the Desired PCA Inventory. The Achieved PCA Inventory consisted of the items on the Desired PCA Inventory slightly reworded so that the respondent could indicate the accuracy of a statement as applied to his present situation. For example, one statement might read, "I take frequent trips away from my neighborhood." The respondent indicated the accuracy of the statement on a six point continuum ranging from Very Inaccurate to Very Accurate. The reliability and validity data reported next were computed on the Desired PCA Inventory only.

Reliability and Validity Checks

Three different types of reliability data were collected. The first reliability analysis sought to confirm the a priori division of the inventory into three distinct scales. A measure of internal scale homogeneity was obtained through the use of the Kuder-Richardson formula (K-R 20). The computed coefficients were .61 for Privacy, .67 for Community and .64 for Activity Space. The next reliability check determined that the obtained scores were accurate reflections of individuals' true scores. A split-half reliability analysis based on the same sample on which the K-R 20 was computed (N=60) yielded Spearman-Brown corrected reliability coefficients of .75, .77 and .65 for Privacy, Community and Activity Space, respectively. The Standard Errors calculated on this sample were 1.2, 1.1 and 1.0 for Privacy, Community and Activity Space with Standard Deviations of 9.2, 4.4 and 7.4, respectively. The possible range of scores on each scale was 10 - 60.

The third type of reliability information determined the stability of an individual's score over time. Approximately three months after the initial inventory was administered, the original respondents completed the inventory for a second time. The test-retest correlation coefficients were .87, .58 and .74 for Privacy, Community and Activity Space, respectively. The Desired PCA Inventory appears to be tapping relatively enduring attitudes that do not change drastically within a short period of time.

Validation data using the peer nomination procedure often employed in validating personality inventories (11) confirmed that the Desired PCA Inventory adequately measured what it was designed to measure. Individuals were asked to read descriptions of six hypothetical persons. These six descriptions presented idealized persons with preferences for very high, or very low, levels of privacy, community and activity space. If the individuals knew of someone who definitely fit one of the descriptions, they were asked to give the Desired PCA Inventory to this person and have him return it in the stamped, self-addressed envelope provided. All responses were confidential, although a special mailbox code on each envelope allowed identification of the category for which the person had been nominated. A t-test analysis revealed that the differences between the means for the "high and low" groups on all three scales were statistically significant. Hence, the Desired PCA Inventory successfully discriminated between individuals categorized by their friends as having low or high preferences for privacy, community or activity space. We now turn to the primary question of this paper--the question of the interrelationships among privacy, community and activity space.

Interrelationships Among the Three PCA Scales

The correlations between the three PCA scales are presented in Table 2. These results are based on two different population samples. The Canadian sample consisted of 60 residents of an urban neighborhood in Ottawa. The British sample consisted of 139 London residents. Again, a representative cross-sample of the population was obtained on the basis of age, sex and type of dwelling. Several words were changed in the British version of the Desired PCA Inventory in order to maintain the same connotations based on different word usages between Britain and North America.

Looking first at the Canadian results, a statistically significant negative correlation was obtained between Desired Privacy and Desired Community (-.25), but not between Achieved Privacy and Achieved Community (-.09). There was no significant relationship between either Desired Privacy and Desired Activity Space (-.03) or Achieved Privacy and Achieved Activity Space (-.19). There was a statistically significant relationship between both Desired Community and Activity Space (.42) and Achieved Community and Achieved Activity Space (.41). The scales were scored such that higher values represented a desire for, or achievement of, more privacy, more community and a localized activity pattern.

TABLE 2

Pearson Product-Moment Correlations Between the Scales on the Desired PCA and Achieved PCA Inventories

Desired PCA (Canadian Sample)	
Desired Privacy with Desired Community	-.25*
Desired Privacy with Desired Activity Space	-.03
Desired Community with Desired Activity Space	.42**
Desired PCA (Britisg Sample)	
Desired Privacy with Desired Community	-.22
Desired Privacy with Desired Activity Space	.03
Desired Community with Desired Activity Space	.16
Achieved PCA (Canadian Sample)	
Achieved Privacy with Achieved Community	-.09
Achieved Privacy with Achieved Activity Space	-.19
Achieved Community with Achieved Activity Space	.41**

*$p < .05$
**$p < .001$

Only the Desired PCA Inventory was administered to the British respondents. The British results were similar to the Canadian results, except for the relationship between Desired Community and Desired Activity Space (See Table 2). The correlation between Desired Community and Desired Activity Space was not statistically significant (.16).

What should we make of these results? In general, one could argue that these results support the proposed model of orthogonality with only slight modifications. Complete independence did not exist between Desired Privacy and Desired Community, but the correlations were low. The correlations between Achieved Privacy and Achieved Community were not significant. Keller's speculation (14) concerning these two concepts was not confirmed. Desired or Achieved Privacy does not seem to be related to any large extent to the spatial aspects of one's activity patterns in the urban environment.

The highest correlations were obtained between Community and Activity Space for the Canadian sample. In this sample, both Desired and Achieved Community were associated with a localized activity pattern. On the basis of the Canadian results, Webber's notion of a "community without propinquity" (33) would not be totally rejected, but it might be somewhat less tenable than previously supposed. However, the British results do not replicate this moderate relationship between Desired Community and Desired Activity Space. Which results most closely reflect reality? Information concerning the social and physical characteristics of the residential areas from which the samples were drawn would aid in clarifying these differences. The neighborhood from which the Canadian sample was drawn would probably be an atypical North American neighborhood. One of the central characteristics is its cultivation of locally based activity patterns. The residents take great pride in the many locally owned shops, the number of places to worship and the opportunities for various types of recreation within the area. Thus, one would expect to find a positive relationship between community and activity space. On the other hand, the British sample was more of an area sample of London at large; hence, the British result might reflect the relationship one would expect to find over an entire urban population. Of course, many other variables may have influenced these results, but if further sampling obtained lower Community/Activity Space correlations, then community, as defined in this project, could be construed to be independent of a geographical area.

These results do not allow a black and white picture to be drawn concerning the interrelationships among privacy, community and activity space. This is probably an accurate reflection of reality, theoretical speculation to the contrary. It would be useful to know which specific behaviors most often contribute to the significant relationships revealed through analysis of the PCA Inventories. The task of collecting the necessary behavioral data still awaits the patient researcher. The purpose of this paper was not to discuss at this time detailed design considerations on the basis of the conceptual analyses of privacy, community and activity space, but simply to caution design specialists to examine their implicit assumptions concerning the interrelationships among these three concepts. The use of an inventory such as the PCA Inventory might prove useful in obtaining information on the explicit relationships among privacy, community and activity space, as viewed by the eventual users of the built environment. The inventory cannot tap all of the nuances associated with these concepts, but it may provide a clearer conceptual framework within which further refinements may be considered.

REFERENCES

1. Altman, I. *The environment and social behavior.* Monterey, California: Brooks Cole, 1975.

2. Bookchin, M. *The limits of the city.* New York: Harper Colophon, 1974.

3. Breckenridge, A. C. *The right to privacy.* Lincoln, Nebraska: University of Nebraska Press, 1970.

4. Brooks, R. O. *New towns and communal values: A case study of Columbia, Maryland.* New York: Praeger Publishers, 1974.

5. Buttimer, A. Social space and the planning of residential areas. *Environment and Behavior,* 1972, *4:* 279-318.

6. Chermayeff, S. and Alexander, C. *Community and privacy.* New York: Doubleday and Company, 1963.

7. Clark, D. The concept of community: A re-examination. *Sociological Review,* 1973, *21:* 397-416.

8. Cooley, C. H. *Social organization.* New York: Charles Scribner's & Sons, 1929.

9. Dentler, R. A. *American community problems.* New York: McGraw-Hill, 1968.

10. Goffman, E. *The presentation of self in everyday life.* New York: Doubleday, 1959.

11. Gough, H. G. *Manual for the California Psychological Inventory.* Palo Alto, California: Consulting Psychologists Press, 1957.

12. Hillary, G. A. Definitions of community: Areas of agreement. *Rural Sociology,* 1955, *20:* 111-123.

13. Jourard, S. *Healthy personality.* New York: Macmillan, 1974.

14. Keller, S. *The urban neighborhood: A sociological perspective.* New York: Random House, 1968.

15. Keller, S. Human communications and social networks at the micro-scale. *Ekistics,* 1970, *179:* 306-308.

16. Kelvin, P. A social psychological examination of privacy. *British Journal of Social and Clinical Psychology*, 1973, *12:* 248-261.

17. Keyes, R. *We, the lonely people: Searching for community*. New York: Harper and Row, 1973.

18. Lansing, J. B., Marans, R. W. and Zehner, R. B. *Planned residential environments*. Ann Arbor, Michigan: Institute for Social Research, 1970.

19. Margulis, S. T. Privacy as a behavioral phenomenon: Coming of age. In S. Margulis (Ed.), *Privacy*. Symposium papers at the meetings of the Environmental Design Research Association, Milwaukee, Wisconsin, 1974.

20. Marshall, N. J. C. *Orientations toward privacy: Environmental and personality components*. Doctoral Dissertation, University of California, Berkeley, California, 1970.

21. Marshall, N. J. C. Privacy and environment. *Human Ecology*, 1972, *1:* 93-110.

22. Maslow, A. H. *Motivation and personality* (2nd ed.). New York: Harper & Row, 1970.

23. Mead, M. Neighborhoods and human needs. *Ekistics*, 1966, *123:* 124-126.

24. Milgram, S. The experience of living in cities. *Science*, 1970, *167:* 1461-1468.

25. Minar, D. and Greer, S. *The concept of community*. Chicago: Aldine, 1969.

26. Poplin, D. E. *Communities: A survey of theories and methods of research*. New York: Macmillan, 1972.

27. Proshansky, H. M., Ittelson, W. H. and Rivlin, L. G. (Eds.). *Environmental psychology: Man and his physical setting*. New York: Holt, Rinehart, and Winston, 1970.

28. Rapoport, A. Some perspectives on human use and organization of space. Paper presented at the meetings of the Australian Association of Social Anthopologists, Melbourne, Australia, 1972.

29. Reubhausen, O. and Brim, O. Privacy and behavioral research. *American Psychologist*, 1966, *21:* 423-437.

30. Slater, P. *The pursuit of loneliness: American culture at the breaking point.* Boston, Massachusetts: Beacon Press, 1970.

31. Slater, P. *Earthwalk.* Garden City, New York: Anchor Press, 1974.

32. Turnbull, A. A., Jr. Privacy, community and activity space: An exploratory investigation. Unpublished Doctoral Dissertation, Carleton University, Ottawa, Canada, 1977.

33. Webber, M. M. Order in diversity: Community without propinquity. In L. Wingo (Ed.), *Cities and space.* Baltimore, Maryland: Johns Hopkins Press, 1963.

34. Wellman, B. The form and function of future communities. In L. S. Bourne, R. D. Mackinnon, J. Seigel and J. W. Simmons, *Urban futures for central Canada: Perspectives on forecasting urban growth and form.* Toronto, Canada: University of Toronto Press, 1974.

35. Wellman, B., Craven, P., Whitaker, M., Stevens, H., Shorter, A., DuToit, S. and Bakker, H. Community ties and support systems: From intimacy to support. In L. S. Bourne, R. D. Mackinnon and J. W. Simmons, *The form of cities in Central Canada.* Toronto, Canada: University of Toronto Press, 1973.

SECTION 2: CROWDING

INTRODUCTORY NOTES

It is no coincidence that the discussion of crowding is the largest of this volume. Crowding is related to, and as hard to define as, concepts of community, privacy and communality. The literature on density, information overload and other aspects of what we call the experience of crowding is vast. From a design point of view, a central consideration may be that crowding interrupts action cycles and, therefore, makes an impact on both feelings of community and privacy by disrupting communications and arousing feelings of stress.

The first paper by Karlin, Epstein and Aiello, researchers from Rutgers University, reports on a field experimental approach. Male and female college freshmen housed in rooms by two's and three's were asked to provide urine samples, perform cognitive tasks and fill out questionnaires and room-use logs. The Cornell Medical Index and the stability of the various living conditions were used to assess the effects of crowding. All subjects in the rooms for three showed increased arousal over time, as measured by cognitive task performance. This is to be expected from laboratory experiments, but, additionally, it was found that females in the rooms for three showed increased physical and psychological problems, as measured on the Cornell Medical Index, and that not one of these rooms proved stable during the time of the experiment. This would appear to be in contradiction to the authors' previous laboratory experiment findings of greater cohesiveness among women under conditions of crowding. Karlin, Epstein and Aiello make a plea for a systems theory approach, which explains these long-term study results without negating their earlier finding from short-term experiments in unnatural environments.

The next paper, by Stokols, Ohlig and Resnick, from the University of California campus at Irvine, also points to the importance of a comprehensive approach in crowding research. It takes Stokols' typology of two dimensions (along which to place the determinants of the crowding experience) as the basis for a long-term study of 27 college students, of which 16 were females. Data were collected from different questionnaires, and a tally of all medical visits on and off campus during the nine-month study period was kept. In support of the neutral/personal thwarting dimension, which predicts that personally significant crowding situations are perceived as more significant, the authors found that assessments of residential crowding correlated with unfavorable reactions to non-residential settings. Additionally, there were significant correlations between ratings of residential crowding, alienation from roommates and perceived social climate of the residence. In support of the primary/secondary environment dimension, which predicts that crowding in primary (e.g., residential) environments has more carry-over effects than crowding in secondary (e.g., commercial) environments, the authors found that students' course grades and their visits to physicians were predicted by subjective ratings of their residential environment. The authors say that it is unclear from their study whether crowding promotes medical problems or whether chronically unhealthy or poor students are more susceptible to crowding experiences. This problem of interpretation of adverse medical effects will also appear in the next paper by Booth. It is clear, however, that use of Stokols' typology, by specifying contextual factors, significantly reduces the number of baffling questions surrounding research into crowding.

The article by Booth is one of an ongoing series of sociological studies by him and others on census tract data of the city of Toronto. As is the case with the next study by Davis, such large-scale studies cannot reveal individual factors in the same fashion as in college student research, but the case numbers are larger. Booth's data from hundreds of families shows a generally good physical and mental adaptation to the levels of crowding commonly experienced in Western cities. One of the findings is that objective and subjective crowding have but a small adverse effect on health, different for men and women and with a possible influence from experienced childhood conditions. For the designer, it is important to know of the differential effects of crowding on the old and the young, the child and the adolescent, etc. Such knowledge may eventually be translated into design criteria for different living situations. Also, although Booth says that humans appear to tolerate rather high levels of congestion, the designer must remember that this relates to statistical averages of the number of people per room and households per block face. Booth's admonition to direct attention and resources toward housing design should, therefore, be interpreted as another way of accentuating the importance of context.

The final paper is a study by Davis of the data from more than 300 interviews with residents of apartments and homes in Queens, New York, and the unobtrusive observations of these people's reaction to strangers. Comparisons were made between the perceived quality of urban life in four neighborhoods with different characteristics, e.g., blacks vs. Jewish, high vs. low neighborhood and dwelling density, short vs. long length of residence, and type of home or apartment. The study is interesting for its innovative observational methodology and its attempt at integrating findings from the earlier papers in this section into its conclusions. The author suggests: "When people believe a situation to be crowded it is real in its physiological consequences, as well as in its sociological consequences." (p. 146)

The papers in this section again imply a plea for non-dogmatic design approach. We are accepting the likelihood that our concepts of crowding are in for fundamental changes and that traditional cause-effect relations are too simplistic. Also, the need to reckon with possibly hidden long-term effects of the crowding experiences should make designers heed the warning of Rene Dubos that man can "become adjusted to conditions and habits that will eventually destroy the values most characteristic of human life" (1, p. 87).

REFERENCES

1. Dubos, R. *Man, Medicine and Environment*. New York: F. A. Praeger, 1968.

STRATEGIES FOR THE INVESTIGATION OF CROWDING[1]

Robert A. Karlin, Yakov M. Epstein, John R. Aiello

Rutgers, The State University of New Jersey

New Brunswick, New Jersey

ABSTRACT

A variety of methods for systematically investigating crowding have merged. These include studies of humans and animals in both laboratory and field settings. Each of these approaches is viewed as having unique strengths and limitations. The potential of these varying research strategies is explored and the need for appropriate caution in generalizing from the results obtained is noted. A recent attempt to study real world crowding in a natural experiment is commented on in detail.

INTRODUCTION

Recently there has been a great deal of interest in the systematic study of the effects of crowding. Researchers have adopted a variety of approaches in the investigation of this phenomenon. Geographers have mapped concentrations of individuals in a variety of settings. Some sociologists have explored the relationship between population density and social pathology, while others have conducted interviews of residents of crowded areas. Some psychologists have studied the effects of density on such variables as task performance, physiology and social behavior. Others have concentrated on studying phenomena associated with personal space looking at nonverbal reactions to inappropriately close distances

[1]This research was supported in part by grants from the Rutgers University Research Council and by Grant #HD-8546-01 from the National Institute of Child Health and Human Development to the authors. All authors contributed equally to this endeavor and the order of authorship was determined solely on a chance basis.

between interactants. Still other behavioral scientists have studied the effects of crowding on animal populations in both laboratory and field situations. It would seem necessary at this time to consider the strengths and limitations of some of these approaches, so that we may better learn what each can contribute to our understanding of how crowding affects human behavior. Finally, we would like to discuss another approach which we feel may further contribute to our understanding of the effects of crowding. Throughout this paper, we will be talking in general terms about the literature in each of these areas rather than referring to specific investigations. We would note that several reviews of the literature have recently appeared. Lawrence (9) and Freedman (6) have reviewed the literature on territoriality, while Evans and Howard (5) have reviewed the literature on personal space. Finally, Altman (2) has surveyed the literature in several of the aforementioned areas. In the main, these reviews have focused on substantive findings and have not considered the strengths and limitations of these disparate research strategies.

Animal Studies

Some of the most dramatic effects of crowding have been found in the animal literature. For example, severe breakdowns in social organization have been observed in experimentally crowded rodent populations. Under conditions of severe crowding, maternal behavior disintegrated, cannibalism occurred, and infant mortality was extremely high. Many male animals displayed heightened aggression and bizarre sexual behavior. The evidence suggests that continued crowding would have completely destroyed the colony. Other animal populations have been studied in the wild. Under conditions of high population density, adverse physiological reactions sometimes leading to dramatically heightened mortality rates have been observed. It should be noted that these animals had an adequate supply of food and were initially healthy. Thus, the adverse effects were clearly attributable to crowding. This research strategy has several obvious strengths. First, the severity of crowding which can be induced in experimental animal populations is much greater than can be created in human populations. Further, the relatively short life span of these animals enables the investigator to observe the effects of crowding not only throughout the entire life cycle, but even across generations. Finally, the ability to sacrifice animals at any point in time has enabled investigators to directly assess the physiological effects of crowding. On the other hand, it must be noted that this approach has some severe limitations. Generalization from animals to humans is always problematical. This concern is especially important when considering reactions to adverse environments. There is a substantial literature dealing with the cognitive mediation of the effects of stress on social behavior and physiological responses. For example, the work of Lazarus and his colleagues points out the importance of cognition in the appraisal of and adaptation to stressful stimuli.[2] Further, the effectiveness of psychotherapy is largely based on our ability to change our reactions to our environment by changing the way in which we think. As a result of this cognitive ability we would expect that in contrast to animals, human beings should be much less adversely affected by environmental stressors such as crowding. Consequently, it may be even more important, in human populations, to look at the costs of continued adaptation rather than the direct effects of adverse environmental stimulation.

Correlational Approach

Let us turn then to some of the human costs resulting from living in crowded environments. A good deal of our knowledge of the effects of crowding on humans is based on the sociological literature which correlates degrees of population density with various indicators of social and physical pathology. These studies have looked at room density, dwelling unit density and neighborhood density and have linked high levels of these densities to such negative outcomes as increased crime, infant mortality, morbidity and problems of mental and physical health. These studies enable us to investigate the effects of real world urban crowding and suggest that crowded environments generally have adverse effects. There are, however, some problems with this research. First, the most crowded urban areas are usually inhabited by people who differ on a variety of dimensions from their less crowded counterparts. They are generally poorer and may be members of ethnic minority groups. They may suffer from malnutrition, have access to less adequate health care services, and have norms and expectations quite different from those of persons living in less densely populated areas. Thus, attempts to evaluate the effects of crowding may be confounded by the difficulty of assessing the contribution of each of these factors. Recent studies have utilized highly sophisticated statistical techniques in an attempt to deal with this problem. However, the success of these techniques is still in question. Issues such as the creation of appropriate indices of socioeconomic status and the choice of cutting points for measures of density have not yet been adequately resolved. For example, two recent studies (Galle, Gove and McPherson, 7; Ward, 12) have used exactly the same data to study the effects of density and social class in Chicago. Both studies found adverse effects in the more densely populated neighborhoods. However, one study seemed to demonstrate that crowding rather than social class was primarily responsible

---

[2] Lazarus, R.: *Psychological Stress and the Coping Process*. New York: McGraw Hill, 1966.

for these effects, while the second study marshalled evidence to indicate that the effects were mostly due to social class. In fact, given the usual relationship between poverty and crowded living conditions, attempts to partial out the effects of each element may only succeed in statistically manipulating measurement error.

Another limitation of this approach is that it focuses on antecedents and consequences of crowding, while failing to consider the processes by which people either successfully or unsuccessfully cope with crowded environments. By looking only at trends over large aggregates, this approach cannot possibly inform us about individual responses to crowded environments. For example, while crowding may be associated with heightened rates of juvenile delinquency, not all children in the neighborhood will turn to crime. Why one youngster is involved in illegal activities and a second lives within the law is not a question that can be answered at this level of analysis. The inability to answer such questions is not a fault of the investigator but is rather an inherent limitation of this technique. One important consequence of this limitation is that while the technique may highlight the scope of the problem, it does not provide us with information about how to ameliorate the effects of crowding. Such information might be more readily available by studying successful and unsuccessful adaptive responses to the environment.

Surveys and Interviews

A different sociological approach utilizes surveys and interviews rather than archival data. A major example of this approach is Mitchell's study of crowding in Hong Kong (10). Mitchell found that most people in the study showed surprisingly few ill effects due to crowding. These results are particularly striking when one considers that residential densities in Hong Kong are among the highest in the world. Mitchell notes that in one government housing project in Hong Kong, density reached 4,100 people per acre. While crowding did cause some differences on relatively minor variables such as the amount of unsupervised play by children, it did not seem to be related to emotional illness, impaired work performance, nor did it have adverse effects on marital relations. Once again, these results must be understood in the proper context. Mitchell was studying the effects of crowding in a culture very different from our own. The literature on personal space shows that there are clear differences among cultures in spatial preferences. Further, there seem to be cultural differences in the cognitive mediation of stress reactions. Simply stated, situations which may be stressful for Americans may not be stressful for Asians. Additionally, while the correlational studies previously mentioned used records of actual behavior, Mitchell relied on the report of his respondents. What prople do and what people say they do is often at variance.

Psychological Experimentation

The next approach to be considered is psychological experimentation. Usually, these studies have involved short-term exposure (anywhere from five minutes to four hours) to very high densities in laboratory settings. Attempts to demonstrate that crowding is an aversive stimulus by measuring its effects on cognitive task performance have met with mixed success. In terms of social behavior, crowding does not seem to have simple effects. Usually, males and females respond differently to crowded settings. Men become more competitive and women more cooperative when crowded. On self-report measures of phenomenological reactions to crowding, people indicate that crowding is unpleasant. Measures of skin conductance seem to indicate that crowding produces arousal. While these studies manage to avoid some of the confounding factors which make the correlational studies difficult to interpret, they have several limitations of their own. Karlin and his colleagues have suggested elsewhere that the key problem here is the setting specific nature of crowding events (8).

Although the label, "crowding," has been applied to a very diverse set of phenomena, the majority of time that urban dwellers spend in crowded environments occurs in a small number of relatively discreet settings. The life of an urban dweller usually involves a fairly limited set of activities. Employed persons get up in the morning, travel on mass transit or in cars to work, stay at their work settings all day long (except for a lunch break), and then return home in the evening. During evenings and weekends they shop and make use of recreational facilities such as parks, restaurants and theatres. On occasions, they may participate in a demonstration or large meeting or otherwise be part of a crowd. We submit that the majority of situations which people label "crowded" occur in one of these contexts.

If it is true that people are crowded in these relatively discreet situations, there are some major implications for the understanding of real world crowding. If we are to understand how crowding affects people and how it may be ameliorated, it seems necessary to phrase such understanding in terms of these discreet contexts. For example, if we are to understand crowding in residences, we must concentrate our research on variables relevant to the study of this context. Crowding people with strangers for a short period of time and studying their task performance while in that setting will have little to tell us about the effects of residential crowding. Many laboratory studies, including our own, have asked such irrelevant questions.

In order for the applied researcher to create conceptual analogues to real world crowding, he must ask a series of questions before setting up his laboratory paradigm. First and foremost the researcher must ask, "What setting do I wish to study?" Here he may choose from among five prototypical settings: residential settings, work settings, mass transit settings, shops and restaurants and crowds. Second, he must ask, "What are the important events which evoke the label of crowding in that setting?" We believe that there are four major events. These are: congestion-resource scarcity, an inability to control and limit interaction with others, close physical proximity to others, and a very large number of interactants in a given setting. Next for the applied researcher whose interests concern the ability to generalize about the effects of crowding outside the laboratory context, the question of what variables are important in real world crowding must arise. This is not the case for the basic researcher who may be interested in the effects of overstimulation, spatial restriction or the presence of large numbers of people on other social psychological processes. One would not, for example, require the student of sensory deprivation to justify his paradigm by pointing to its real world analogue. However, the applied researcher must adequately conceptualize both his independent and dependent variables in light of what actually occurs. Moreover, he must make a distinction between logical possibilities and applied research priorities. While it might be interesting to know how the crowded subway rider feels about the person next to him, from an applied standpoint, it is probably more important to assess the effects of the ride on his subsequent interactions with family members at home. Similarly, in terms of effective interventions, real world considerations may be taken into account. If one possible intervention in terms of crowded subways is to provide much more space by building many additional subway cars, practical considerations render this a low probability occurrence. Alternate strategies such as exploring such social manipulations as traveling in the company of friends and thereby reducing the salience of the crowded environment may be a more profitable investment of such research efforts.

Laboratory studies of crowding to date have not, for the most part, had relevant applied considerations in mind. We have still to create appropriate laboratory analogues to real world situations in which people become crowded. Finally, let us consider a different sort of experimentation. Projective techniques have been used to ascertain reactions to crowding and certain elements of design. Usually, research subjects are asked to place objects representing people in scale model rooms in order to determine the point at which they judge the environment to be crowded. On the face of it, this approach would seem to have a great deal of merit for those concerned with design questions. This is especially true since data can be gathered with little time and effort. However, for a series of theoretical reasons which we have previously discussed elsewhere (1), such studies are likely to invoke very different judgmental processes than the judgment of crowding resulting from active participation in settings differing in density. This position has recently received strong support from a series of studies which showed practically no relationship between projective and actual measures of interpersonal distance.

In summary, each of the above approaches has clear strengths and limitations. Further, we would note that none of these strategies allow us to examine the processes by which people cope with crowded living conditions.

A Field Experimental Approach

In this regard, we would like to describe an approach which we have recently employed to investigate the effects of crowding in the dormitories at Rutgers University.

For several years, Rutgers University has been experiencing a shortage of student housing. With the addition of new colleges, enrollments have increased rapidly. The shortage of state budgeted funds has meant that the University has been unable to construct new housing facilities. The poor housing condition in the New Brunswick area, the high cost of such housing and the generally high crime rate in the area have combined to create a great demand for on-campus housing. All these factors have contributed to the crowded conditions in the dormitories. Different undergraduate colleges at Rutgers have evolved differing solutions to this problem. At one of the undergraduate colleges, for example, a practice of "staging" students has been used. This means that 18-20 students are placed in barrack-like lounges. The only facilities available to these students are beds. In this situation, students wait until rooms become vacant--either because other students do not arrive for the beginning of the semester, or because they move out of rooms or leave school. The staging arrangement usually does not last for more than several weeks. At another undergraduate college, however, a different procedure is being used. Rather than staging some students in a large room and allowing all other students to live two to a room, this college has decided to triple up students in rooms that were intended for two-person use. Since there are fewer students than would be required to triple up all the rooms, some rooms are tripled and some are not. The procedure for deciding who will be tripled is left to a chance lottery. This situation created the conditions for an ideal "natural experiment." There was a population of males who had been tripled, females who had been tripled, and their counterparts of both sexes who were living two to a room. All these people came from relatively homogeneous backgrounds with respect to socioeconomic status. They had equivalent educational backgrounds (all were college freshmen), were about equally healthy when they arrived at college, and had all been randomly assigned to living conditions. This provided an opportunity to study the effects of crowding from a longitudinal perspective in a manner uncontaminated by socioeconomic class and health factors which have usually differentiated persons who live in crowded urban conditions from those who do not.[3] At the same time, it provided some of the strengths of the experimental method and an opportunity to examine the processes by which people cope with conditions of crowded living.

METHOD

Subjects

A sample of 31 rooms (7 tripled male rooms, 7 tripled female rooms, 7 doubled male rooms and 10 doubled female rooms) was randomly selected from the available population. Potential subjects were offered an incentive of $25.00 each to contribute about 15 hours of their time during the course of the semester to the research project. All but one person so approached agreed readily to participate in the study. Subjects were told that the study would investigate patterns of adjustment of college freshmen to college and dormitory life.

Dependent Measures

The present research into long-term crowding was greatly influenced by Irwin Altman's model of contact regulation (2). The Altman model provided a guiding framework for our choice of dependent variables. Since the model is comprehensive, dealing with a wide range of variables, an attempt was made to collect information on as many of them as possible. Analyses have been completed on only those measures which directly relate to arousal or are clearly

---

[3]It is clear that in addition to experiencing restricted space and a scarcity of resources, tripled Ss lived in triadic as opposed to dyadic groups. These configurations have properties of their own, such as the presence of coalitions in the triads which would not be possible in two-person situations. One of our students is presently conducting a doctoral dissertation on the effects of group size as opposed to spatial factors in the present study. (M. Geller: The Behavioral Impact of Group Size and Available Space: An Analysis of Dyadic and Triadic Roommate Arrangements.)

pertinent to the extension of laboratory findings of sex differences in response to crowding.

Two sets of measures were used to determine whether crowding was arousing and stressful. The first involved the measurement of unbound cortisol obtained from urine samples. The presence of long-term stress should cause heightened levels of cortisol to be secreted. The procedure for assaying cortisol levels is extremely costly and complex; because of the expense of this procedure, a subsample of five subjects in each of the conditions was randomly selected for this phase of the study. From each of these subjects 24-hour urine samples were collected twice during the course of the semester at monthly intervals. The second set of measures involved cognitive performance. Simple and complex cognitive tasks were used to measure arousal. The simple task consisted of crossing out the letter "A" every time it appeared in a list of words. The number of A's crossed out in a two-minute interval served as the dependent variable. The complex task consisted of a set of nonsense syllogisms whose truth value had to be ascertained by the subject. The number of correctly judged deductions in a four-minute time period was the measure of complex performance. Two equivalent forms of each of these tasks were created and administered early and late in the semester. Arousal would be indicated by increases in simple task performance and decreases in complex performance relative to baseline data.

Two major indices were used to assess the effects of crowding. The Cornell Medical Index was used to assess the effects of crowding on health. Clearly, an increase in psychological and physical illness would indicate some detrimental effects of crowding. Second, we observed the stability of the various living conditions. A variety of opportunities arose during the course of the semester for students to leave their original rooms. Since some students left school or decided to commute during the semester, alternate living arrangements could be found. Moving out of one's room should be a clear indication of dissatisfaction with living conditions. Finally, a series of questionnaires and a Room Use Log elucidated the process of adjusting to crowding.

Checks on the Manipulation

While it was clear that the dormitory rooms were built to house only two people, it could not be assumed that three persons occupying these rooms would consider themselves to be crowded. It was possible that the small size of these rooms would cause even two people to feel crowded and the addition of a third person might have made no difference. Alternatively, it was possible that even three people in the room might not have felt particularly crowded--clearly, they had more space than did subjects in laboratory studies

TABLE 1

Feeling of room crowdedness by male and female doubles and triples*

	Doubles	Triples
Female	2.5	6.1
Male	3.2	5.1

*1 = not crowded; 7 = very crowded

TABLE 2

Satisfaction of doubled and tripled males and females*

	Doubles	Triples
Female	2.0	5.0
Male	2.3	4.0

*Lower scores are associated with greater satisfaction.

of crowding. Two questions were therefore asked tapping perceived crowding and satisfaction with living conditions. Subjects in tripled rooms perceived themselves as significantly more crowded than did persons living in double rooms ($F=67.9$, $df-1/58$, $p < .001$). Interestingly, while both men and women in tripled rooms differed significantly from their less crowded counterparts, the effect was stronger for women that it was for men ($F=6.0$, $df=1/58$, $p < .025$). Examination of reported satisfaction with living conditions reveals the same pattern. Tripled subjects were significantly less satisfied than were doubled subjects ($F=61.3$, $df=1/58$, $p < .001$). The effect is stronger for crowded women than it is for crowded men ($F=5.93$, $df=1/58$, $p < .025$).

Arousal

Two sets of measures were used to study arousal: cortisol levels and cognitive task performance. Cortisol levels did not show significant differences. However, tripled subjects showed an increase over time. Large individual differences in cortisol level as well as the small size of the sample may have prevented these differences from being clearer. Cognitive task performance, on the other hand, revealed clear indication of arousal over time. Subjects in double rooms showed improved performance over time on both

TABLE 3

Cortisol levels of doubled and tripled students

	Time 1	Time 2
2 person	41.8	34.7
3 person	44.5	47.6

TABLE 4

Cognitive performance of doubled and tripled students

Triples			Doubles		
Number Who:	Simple	Complex	Number Who:	Simple	Complex
Increase	22	8	Increase	16	13
Decrease	7	19	Decrease	6	8

the simple and the complex tasks ($X^2=0.57$, df=1, n.s.). Tripled subjects, however, showed the expected pattern of arousal, i.e., improved performance on a complex task ($X^2=12.02$, df=1, $p < .001$).

Health and Room Stability

Laboratory findings have generally demonstrated that the major detrimental effects of crowding occur for male groups. The two measures to be reported below reveal that in this instance, the major detrimental effects occurred for crowded women. The Cornell Medical Index revealed that crowded women have more physical and psychological problems than did the other three groups, whereas male tripled subjects did not differ from males in doubled rooms ($F=7.2$, df=1/58, $p < .01$). Looking at the relative stability of the various living conditions, the same pattern emerges. Tripled rooms were significantly more unstable than doubled rooms ($X^2=9.38$, df=1, $p < .01$). However, this effect was almost entirely caused by the dissolution of tripled female rooms. In all seven cases at least one of the tripled females had left the room for an alternate living arrangement by the end of the semester. This was the case for only two of the tripled male rooms. Thus, tripled female rooms were significantly less stable living arrangements than were tripled male rooms ($p< .01$, Fischer Exact Test). Anecdotally, it is interesting to note that one of the two tripled male rooms which dissolved invited a new third person to join them.

TABLE 5

Physical and psychological problems reported on the Cornell Medical Index by male and female doubles and triples*

	Doubles	Triples
Female	1.93	1.89
Male	1.93	1.95

*The lower the score, the greater the reported problems.

TABLE 6

Stability of double and triple rooms

	Double	Triple
Broke up	1	9
Stayed together	16	5

TABLE 7

Stability of female and male triples

	Female	Male
Broke up	7	2
Stayed together	0	5

DISCUSSION

These results would seem to indicate that there are a number of differences in the effects of the environment on students living in tripled rather than in doubled room arrangements.[4] Tripled as opposed to doubled subjects saw themselves as more crowded and were less satisfied than doubled students. They seemed to be more aroused. However, when considering the effects of crowding on health and the stability of living arrangements, it is the crowded women who are most negatively affected. It should be noted that the same pattern emerged on the manipulation checks where the effects were stronger for women. While the data on arousal were consistent with results obtained in the laboratory, it would seem that on the face of it, the sex effects have been reversed.

The most obvious difference between this study and the laboratory studies is that this study concerns long-term crowding, while only short-term crowding was studied in the laboratory. In addition, since crowding is stressful, the key mechanism in question is a process of adaptation to stress. Coping mechanisms which work well in relation to short-term situations may prove to be extremely maladaptive in long-term situations. This idea is not new; Selye's

---

[4]The study of residential crowding in the Rutgers dormitory was concluded in early 1975. Since that time, the study by M. Geller, mentioned in Footnote 3, has been completed, and a follow-up study by L. Rosen is in progress. These studies have indicated several things. First, the study by Geller indicated that the stress of the triadic living arrangement was due, in large measure, to the coalitional properties of the three-person group. It should also be noted that A. Baum (personal communication) has found that the negative effects of crowded tripled dormitory living seem to be most pronounced for students who are excluded from a coalition between the other two roommates.

The study by Rosen, which was a two-year follow-up on the subjects of the original dormitory population, indicated that three-person female groups were more satisfied with college life than were any other groups after their freshman year. The data analysis for Rosen's study has not yet been completed, and it is not clear why tripled women subsequently showed better adjustment to college life than did persons in any of the other groups. The effect does not seem to be an instance of regression to the mean. We are considering three major possibilities to account for this finding. First, it may be due to the presence of helpful social networks that were formed in the first year. Second, the adverse experience in the first year may have taught these women coping skills which they later used. Third, since the clearest differences in Rosen's study emerged on self-report measures, the data may be due to social judgment considerations. Since tripled women were most negatively affected in their freshman year, the contrasted improvement brought about by the change in living conditions may have seemed more striking to them than to members of other groups.

TABLE 8

Proportions of students in room

	Male	Female
In room	19	32
Out of room	16	9

notion that the very process of adaptation to stress may lead to further costs resulting from the efforts to adapt is the classic formulation of this notion. In the laboratory, it has been seen that short-term crowding leads to increased interdependence among women and solitary activity among men. These patterns have been previously identified as a "positive" reaction among women and a "negative" reaction among men. In part, these labels reflect particular values and measures used to study the effects of crowding in the laboratory. For example, crowded women were more cohesive, viewed each other as more similar, and encouraged each other to share their distress (4). It was assumed that these behaviors were "good." In fact, they represent a particular style in coping with stress, i.e., increased interdependence. This affiliative style is typical of women and is also used by them in the long-term situation. It seems quite possible that it is this very process of high levels of interaction and interdependence among women which produced the strong negative effects of crowding on the tripled women. A variety of measures in the present investigation support the view that women employ this high interaction, interdependent style in long-term living situations, while men do not. This is most clearly seen when one examines which subjects spent the most time in their rooms and, hence, with each other. The procedure of questionnaire distribution permitted verification of the notion that women spent more time in their rooms than did men. Questionnaires were delivered to subjects in their rooms. If the subject was not in his or her room, the questionnaire was stamped with a notation requesting him to place it in a locked box on the front desk in the dormitory. In this way, it was possible to ascertain who were in their rooms and who were not. These data clearly indicate that women were in their rooms more often than men (X^2=4.83, df=1, $p < .05$).

In addition, subjects were asked to keep a room-use log for one week's time. On this log, subjects indicated the activities that were conducted in the room and noted the time of the day that they entered and left the room. As can be seen in Table 9, women tended to spend more time each day in their rooms than did men

TABLE 9

Mean number of hours/day spent in own dorm room

	Male	Female
Doubles	5.6	6.1
Triples	5.4	7.0

($t=2.04$, $df=18$, $p < .06$). Parenthetically, it should be noted that only about one hour's time in each of these conditions represented time subjects spent alone in the room. Given that crowding is stressful, this implies that the tripled women were constantly being exposed to an aversive stressful environment. It is not surprising that this took its toll, both in terms of health and room stability. Men, on the other hand, spent less time in their rooms. For them, the exposure to the situation was somewhat more limited. Therefore, the effects of crowding on the men were somewhat mitigated. College environments for freshmen are often stressful in a number of ways. Clearly, any effects of the physical environment are above and beyond these stresses. Often the college dormitory can be a place to retreat from the stresses of college life. This may be particularly true for women, who, from the outset, plan to and subsequently do invest more of their physical and psychological resources in making their room into a home for themselves than do men ($F=11.57$, $df=1/60$, $p < .005$).

Crowded women confront a difficult situation; they wish to retreat into their rooms, but if they do so, they suffer from additional stress. Note that this does not imply that the effects on health and living stability which have been observed in this study were the effects of crowding. Rather, when the stress of crowding is experienced in an otherwise highly stressful situation, its clear negative consequences can be observed.

TABLE 10

Investment of physical and psychological resources

	Male	Female
Intended investment	1.87	1.71
Actual investment	1.92	1.83

CONCLUSION

In his discussion of scientific methods in the behavioral sciences, A. Kaplan (*The Conduct of Inquiry*. San Francisco: Freeman, 1964) describes an incident in which a policeman notices a drunken citizen stooped on all fours and scratching around on the sidewalk under a lamppost. When asked by the policeman what he is doing, the drunkard replies that he is searching for a lost house-key. The policeman asks him where he lives and he points to a dwelling some 50 yards down the block. The policeman asks him whether he remembers losing the key in the vicinity of the lamp-post. "Oh, no," replies the man, "I lost it in the front of my house." "Why, then," asks the policeman, "are you looking for it in front of the lamppost?" "Oh," replies the drunkard, "because that's where the light is." Perhaps there is a lesson in this tale for investigators studying the effects of the environment. On the one hand, simple causal models of the type which are usually amenable to controlled laboratory investigations may not adequately account for phenomena such as crowding. Altman's model of crowding is an example of an alternative to the simple causal model. His approach seems rooted in the systems theory tradition. Clearly, no single laboratory experiment could adequately test his model. Rather, many different investigations are needed to accumulate information which can be seen as consistent or inconsistent with his ideas.

But, in addition to the conceptual issues noted above, one must ask whether the laboratory, as we have been using it, is really the best place to study environment effects? Singer, et al. (11) have made this point in a recent article on commutation stress:

> As psychologists turn to considerations of their environment, it is all important that studies be done of the environment itself. Theoretical or laboratory work can provide theories, hunches, and a host of detailed studies concerning the effects of selected and isolated variables; but they cannot substitute for a study of its aspects of the work outside the laboratory to which they are addressed. It thus requires acts of faith and extrapolation to move from a laboratory phenomena to recommendations of real application. (p. 18)

There is something very curious about our tenacious hold upon traditional laboratory experimentation in order to feel secure in the knowledge we obtain. We submit that when it comes right down to it, the utility of research results for understanding real world phenomena requires an act of faith. In the case of well controlled laboratory experiments, this means on the one hand discounting such factors as demand characteristics and other such potential artifacts and on the other hand making the huge inductive leap between a study manipulating a limited number of variables in an artificial laboratory and a more complex real world phenomenon. Alternatively, when one conducts field studies and cannot control extraneous environmental variables, a different act of faith is required. In the long run, both of these approaches will yield spurious results which will have to await newer and improved studies for a better, though still incorrect, approximation to knowledge.

REFERENCES

1. Aiello, J., Epstein, Y., and Karlin, R. Effects of crowding on electrodermal activity. *Sociological Symposium*, 1975, *14:* 43-57.

2. Altman, I. *The environment and social behavior: privacy, personal space, territory, and crowding.* Monterey, Ca.: Brooks/Cole, 1975.

3. Edney, J. Human territoriality. *Psychological Bulletin*, 1974, *81:* 959-975.

4. Epstein, Y and Karlin, R. Effects of acute experimental crowding. *Journal of Applied Social Psychology*, 1975, *5:* 34-53.

5. Evans, G. and Howard, R. Personal space. *Psychological Bulletin*, 1973, *80:* 334-344.

6. Freedman, J. The effects of population density on humans. In J. T. Fawcett (Ed.), *Psychological perspectives on population.* New York: Basic Books, 1972.

7. Galle, O., Gove, W., and McPherson, J. Population density and pathology: What are the relations for man? *Science*, 1972, *176:* 26-30.

8. Karlin, R., Epstein, Y., Aiello, J. The setting specific nature of crowding. In A. Baum and Y. Epstein (Eds.), *Human response to crowding.* Hillsdale, N.J.: Lawrence Earlbaum Associates, 1976.

9. Lawrence, J. Science and sentiment: overview of research on crowding and human behavior. *Psychological Bulletin*, 1973, *81:* 712-721.

10. Mitchell, R.E. Some implications of high density housing. *American Sociological Review*, 1971, *36:* 18-29.

11. Singer, J., Lundberg, U., and Frankenhaeuser, M. Stress on the train: a study of urban commuting. In A. Baum (Ed.), *Advances in environmental psychology*, Volume 1 (in press). Hillsdale, N.J.: Lawrence Earlbaum Associates, 1976.

12. Ward, S.K. Methodological considerations in the study of population density and social pathology. *Human Ecology*, 1975, *3:* 275-286.

PERCEPTION OF RESIDENTIAL CROWDING, CLASSROOM EXPERIENCES, AND STUDENT HEALTH

Daniel Stokols, Walter Ohlig, and Susan M. Resnik

University of California, Irvine

Irvine, California 92664

A core concern of environmental design research is the impact of the physical and social environment on human health and behavior. Professional designers often approach this issue through direct observation of the relationships between architectural variables and behavioral patterns. Design-oriented behavioral scientists, while sharing designers' concern with the direct linkages between objective environments and overt behaviors, more commonly approach the environment-behavior interface by way of theoretical, or intervening constructs. These constructs help to specify the social and physical conditions under which specific features of the physical environment might correlate with various types of behavior.

The design-practitioner and behavioral-scientist perspectives are, of course, complementary in the sense that the first keeps environmental-design research firmly tied to concrete, measurable variables, while the second assists the researcher in predicting, *a priori*, those environmental dimensions that will exert the greatest influence on behavior across diverse situations. (cf. 1).

The utility of combining design and behavioral perspectives can be illustrated through a consideration of the issue of human crowding. An architecturally-oriented analysis of crowding would focus primarily on physical variables, such as the amount and arrangement of space, and the correlations between these variables and behavioral patterns within specific settings. A typical assumption underlying this approach is that spatial limitation is associated with a number of negative behavioral effects, including withdrawal from social interaction, impairment of task performance, and pathological behavior. The problem with a purely physicalistic perspective on crowding is that it does not account for numerous situations in which limited space is associated with positive rather than negative behavioral consequences; for example, work situations in which the proximity of others enhances collective task performance; or crowded parties which promote camaraderie among members of a group. In order to predict where spatial limitation will induce physiological, psychological and behavioral problems, it becomes necessary to consider the social-psychological dimensions of various situations.

A behavioral science approach to crowding attempt to identify social-psychological factors which mediate the impact of architectural (especially spatial) variables on behavior. A central assumption underlying this approach is that the categorization of environments in terms of both physical and social-psychological dimensions should provide a basis for developing design guidelines pertaining to several related questions, such as: In what types of environments will spatial limitation lead to major disruptions in individual and interpersonal activities? What kinds of adaptive strategies are available to occupants of high-density settings? To what extent will psychological and behavioral deficits associated with crowding in one setting generalize to other situations?

These questions are examined below in relation to a typology of crowding experiences (16). The typology focuses on the subjective experience of crowding rather than on conditions of high-density which may or may not be related to perceived crowding. A number of derivative hypotheses pertaining to the intensity, persistence, and reducibility of crowding experiences are discussed, and some preliminary research on the generalizability of crowding experiences from one situation to another is presented. It is assumed throughout this discussion that a refinement of the crowding construct will provide a basis for predicting crowding potentials and related behavioral impairments as a function of both physical and social dimensions of the environment; and that such prediction ultimately will contribute to the design of physical settings which are maximally congruent with the needs of their users.

A Typology of Crowding Experiences

The typology is based on the distinction between density, a physical condition of limited space, and crowding, a subjective experience of psychological stress in which one's demand for space exceeds the available supply (13). A basic assumption relating to this distinction is that increased demand for space can arise not only in response to direct spatial restriction, but also as a result of social circumstances which sensitize the individual to potential problems posed by continued proximity with others.

Recent analyses have suggested a variety of non-spatial antecedents of crowding. Stimulus overload models, for example, posit that the experience of crowding is heightened by excessive social stimulation (cf. 2, 4, 6, 17, 19). Behavioral constraint formulations link the perception of crowding to restraints on behavioral freedom and infringements on privacy imposed by the proximity of others (cf. 8, 14). And ecological perspectives on crowding suggest that increased demand for space may result from a scarcity of social and/or physical resources in the setting (cf. Note 4; 18).

Stimulus overload, behavioral constraint, and ecological theories of crowding converge on the assumptions that (1) crowding involves the perception of insufficient control over the environment, and (2) increases the desire to put more space between oneself and others as a means of avoiding actual or anticipated interferences. The joint utility of these analyses is that they provide insights into the nature and determinants of perceived crowding. Their major limitation is that they offer few clues concerning the parameters of crowding intensity and persistence. The conditions under which overstimulation, behavioral constraints, and resource scarcities lead to the most disruptive experiences of crowding remain unspecified.

An additional assumption concerning the nature of crowding is required to permit an identification of factors that mediate the intensity and persistence of crowding experiences, *viz.*: Increased demand for space will be most intense, persistent, and difficult to resolve when it is associated with perceived threats to physical or psychological security. Proximity with dangerous or insulting persons, for example, would lead to more intense crowding than the same degree of proximity with others who are seen as posing no threat to the individual's security.

The present typology of crowding experiences incorporates two dimensions which help to "sort out" the determinants of crowding intensity and persistence: neutral/personal thwartings and primary/secondary environments. The thwarting dimension pertains to the nature of interferences imposed by proximity with others. Neutral thwartings are essentially unintentional annoyances stemming from either the social or nonsocial environment, whereas personal thwartings are those interferences intentionally imposed on the individual by other persons. Under conditions of *neutral crowding*, the need for more space relates primarily to physical concerns such as the restriction of movement and the discomforts associated with high-density conditions. To escape these inconveniences, the individual desires more control of the physical, i.e., spatial, environment. In situations of *personal crowding*, increased demand for space relates to both physical and social concerns. Here, the salience of physical inconveniences is increased by the presence of hostile or unpredictable others. To resolve feelings of crowding, the individual must gain control over social as well as physical aspects of the environment.

The primary/secondary dimension of the model concerns the continuity of social encounters in a particular setting, the psychological centrality of behavioral functions performed within the setting, and the degree to which social relations occur on a personal or anonymous level. *Primary environments* (e.g., residential and work environments) are those in which an individual spends much time, relates to others on a personal basis, and engages in personally-important activities. *Secondary environments* (e.g., transportation and commercial environments) are those in which one's encounters with others are relatively transitory, anonymous, and inconsequential.

The model of crowding experiences outlined above suggests several hypotheses. First, it can be predicted that experiences of personal crowding will be more intense, persistent, and difficult to resolve than those of neutral crowding, since the former are more likely to involve perceived threats to one's physical safety or self-esteem, and to induce frustration of expectancies regarding the adequacy of space. Furthermore, assuming that the individual is confined to the situation, perceptual and cognitive modes of adaptation to crowding (e.g., ignoring spatial constraints, adopting more favorable attitudes toward other occupants of the area) will be less viable in situations of personal crowding, because the potential for social conflict is greater there than in instances of neutral crowding.

A second hypothesis is that crowding experiences will be of greater intensity and duration in primary environments than in secondary ones. This prediction is based on the assumption that primary settings tend to be associated with higher expectations of personal control along a greater diversity of need dimensions and, therefore, proximity-related interferences will be more likely to thwart personally-important needs and goals in primary vis-a-vis secondary environments.

A third hypothesis pertains to the generalizability of crowding experiences from one setting to another. It is expected that personal crowding experiences, particularly in the context of a primary environment, will generalize more readily to other situations than will neutral crowding experiences. (The generalizability of crowding experiences could be measured, for example, in terms of a person's need for space, susceptibility to performance deficits, and medical complaints across a variety of settings, subsequent to his/her experience of crowding within a particular situation.) The main assumption underlying this prediction is that personal crowding, because it typically involves ambivalent or negative attitudes toward others, provides a cognitive base from which anxieties about proximity with certain persons in the setting can generalize to other people in different situations.[1] In contrast, the impact of neutral crowding experiences, which are less closely associated with persisting attitudinal changes, would be more confined to the immediate situation.

The present research provides preliminary data which pertain directly to the third hypothesis, and are of general relevance to the first and second predictions, mentioned above. The data were gathered through a three-part campus survey in which college students evaluated their residential environments (e.g., dormitory suites, apartments) during the second and third weeks of the academic quarter, and rated the amenity of a particular classroom setting during the fourth week of the quarter. Subsequently, most of the students who had completed the dormitory and classroom questionnaires consented to have the Student Health Center release information regarding the number of visits they made there during the academic year. They also reported the number of times they had consulted off-campus physicians during the year.

The residential and classroom questionnaires incorporated items pertaining to both physical and social features of the environment. On the basis of questionnaire responses, an attempt was made to predict sensitivity to crowding in the classroom as a function of subjects' ratings of their residential environment. Moreover, the classroom responses of students reporting neutral and personal crowding experiences in their residences were compared in order to detect possible differences in the generalizability of crowding sensitivity from residential to classroom settings. Grades achieved by the students were recorded to determine possible effects of crowding on classroom performance.

The specific predictions of the study were as follows: (1) In a stepwise multiple regressional analysis, the prediction of classroom crowding will be significantly more reliable when ratings of both the physical and social dimensions of the residential environment are employed as predictor variables, than when assessments of the physical environment are used alone. (2) Subjects whose residential evaluations indicate a pattern of personal crowding will express greater feelings of crowding in the classroom than those whose dormitory or apartment ratings reflect the experience of neutral crowding. (3) Students' visits to health centers on and off campus during the academic year, and their course grades during the Fall Quarter, will be significantly associated with their ratings of residential crowding and their evalutions of the physical and social conditions within their residences, as reported by them during the Fall Quarter.

METHOD

Subjects

Participants in the study were drawn from a large undergraduate course offered at the University of California, Irvine, during the Fall Quarter. The enrollment for the course was approximately 400 students. From a listing of students taking the course, prospective subjects were identified on the basis of two main considerations: (1) place of residence and (2) year in college. An attempt was made to comprise the sample primarily of first-year students living in suite-design dormitories, so as to minimize subjects' prior exposure to classrooms on the Irvine campus and to control for variations in residential density and design. Participants were drawn from the same lecture class to control for the effects of professorial style, course content, and classroom design on the data.

From an original listing of 60 prospective subjects, only 21 volunteered to participate in the study and completed both the residential and classroom questionnaires. In view of the small sample size, an additional group of second-year students, some of whom resided in off-campus apartments, was included in the sample. The sample employed in the statistical analyses of the residential and classroom data consisted of 20 females and 11 males.

Near the end of the Spring Quarter, 1975, an attempt was made to contact these same subjects in their dormitory suites or by telephone. Two of the subjects could not be traced and two others did not consent to the release of information regarding their visits to the Student Health Center. This brought the sample size used in the prediction of doctors' visits to 27 subjects. Of these, twelve females and six males resided in double rooms of suite-type dormitories, one female occupied a single dormitory room and three females and five males lived in off-campus residences. Three subjects had changed rooms within the dormitories since the Fall Quarter and one subject had changed his off-campus apartment.

Procedure

Prospective subjects were contacted by phone during the second week of the quarter and informed about a "study of the reactions of college students to dormitory (or off-campus) living conditions at U.C. Irvine." They were told that if they agreed to participate in the survey, they would fill out questionnaires concerning the physical and social attributes of their current residence. The caller further explained that the questionnaire session would last for approximately 20 minutes, and that participants would be paid $2.00 at the session. Students agreeing to participate in the study were asked to report to an office on campus during the second or third week of the quarter where they completed the residential questionnaire.

The collection of classroom data occurred during the fourth week of the quarter. The professor[2] distributed a questionnaire to all members of the class during the initial portion of a class period. The questionnaire was described as part of a research project being conducted by a faculty associate, the purpose of which was to learn more about students' reactions to different kinds of courses at U.C. Irvine. It was emphasized that completion of the questionnaire was entirely voluntary, and that each student's responses on it would have nothing to do with the grade he or she received in the course. Students volunteering to assist in the project were asked to put their student identification number on all pages of the questionnaire to facilitate collation of the data. The professor also noted that questionnaire responses would remain anonymous and would be coded on computer cards for statistical use only.

The questionnaire contained several items regarding the relative advantages and disadvantages of large lecture classes in comparison with other types of courses; small seminars and large discussion classes, for example. Embedded among these items was a set of bi-polar scales concerning the physical conditions and social climate of the present classroom. These were identical to the items included in the dormitory questionnaire.

When the students were approached again during the Spring Quarter, they answered several questions pertaining to their health and were asked to sign a consent form allowing the Student Health Center to release information regarding the number of visits made by each student during the three quarters of the academic year.

Measures and Analyses

The residential questionnaire incoporated six sets of seven-point semantic differential scales. The first included eight items pertaining to subjects' perception of their dormitory suite or apartment in terms of the quality of its physical dimensions. Subjects were asked to indicate the degree to which their residence permitted privacy and was pleasant, comfortable, spacious, quiet, large, uncluttered, and cheerful. The second set of scales related to the perceived quality of social relationships existing among themselves and their suitemates (roommates). Subjects were asked to rate the degree of trust, competition, alienation, similarity, and hostility felt among themselves and other occupants of the residence, as well as the extent to which they tried to make each other feel secure, were considerate of each other's feelings, and confided in each other about personal problems. A third set of scales pertained to the degree of crowding and spatial restriction felt by subjects in their dormitory suite or apartment.

Three other sets of semantic differential scales were included in the questionnaire. Two contained items from Rotter's Internal-External Scale (10) and Keniston's (Note 5) Short Alienation Scales, respectively. On the remaining set, subjects were asked to rate "people in general" along ten dimensions including, for example, "harmful-beneficial," "hostile-friendly," "disturbing-calming," and "bad-good."

Two final assessments were incorporated into the residential questionnaire. First, subjects were asked to draw a map of their dormitory suite or apartment on a blank piece of paper. Second, they were requested to complete the Comfortable Interpersonal Distance Scale (CID), a paper-and-pencil measure of personal space needs (3).

The classroom questionnaire incorporated three sets of seven-point scales. On the first set of scales, subjects were asked to rate how crowded, restricted, threatened, insecure and tense they "usually feel in the present classroom." The second and third item-clusters pertained to subjects' evaluation of the classroom in terms of its physical conditions and social atmosphere, respectively. These items tapped the same dimensions of environmental amenity as those reflected in comparable scales of the dormitory questionnaire. Finally, a number of open-ended filler items were included in the classroom questionnaire which required subjects to list the relative advantages and disadvantages of large versus small- and medium-sized classrooms.

The Spring questionnaire consisted of questions pertaining to the number of times the students had visited medical doctors on or off-campus during the academic year. The Student Health Center located on the University of California campus reported the actual number of visits made by each student during the three quarter terms.

Three major analyses were performed on the data. First, the correlations between residential and classroom assessments of environmental quality were examined. The units of analysis were item-cluster total scores which were computed by summing subjects' responses on the various scales within a particular cluster. Thus, each of the three cluster scores obtained from the residential data was correlated with each of the three total scores derived from the classroom data. Also, the correlations between environmental evaluation scores and the additional indices included in the residential questionnaire were examined.

Second, a series of step-wise regression analyses were performed on the dormitory, classroom and health-center data. In the primary analysis, the classroom measure of felt crowding was employed as the response variable, while cluster scores pertaining to the physical quality of the residential environment, its social atmosphere, and the perception of residential crowding was utilized as predictor variables. In a subsequent analysis, assessments of subjects' internal-externality, chronic alienation, interpersonal-distance preferences, and residential map size were incorporated into the regression equation as predictor variables.

In two other analyses, course grades and visits to the Student Health Center during the year were used as dependent variables. In both analyses, residential crowding, physical conditions and social atmosphere were employed as predictor variables. The classroom atmosphere index was employed as an additional independent variable in the analysis of course grades.

Third, multivariate analyses of variance (MANOVA) were performed on the classroom data, utilizing the major indices of residential evaluation as blocking factors. In the first analysis, subjects were divided into "low crowding" and "high crowding" groups through a median split of their residential crowding total-scores. For the second analysis, subjects' evaluation of the physical environment (EPE) and social environment(ESE) were utilized jointly to distinguish among four different groups of respondents: (1) those who evaluated both the physical and social dimensions of their residence positively (+EPE/+ESE); (2) those who evaluated both dimensions negatively (-EPE/-ESE); (3) those who reacted positively to the physical conditions of their dormitory suite or apartment, but negatively to its social climate (+EPE/-ESE); and (4) those who reacted negatively to the physical conditions, and positively to the social climate of their residential environment (-EPE/+ESE). On the basis of these groupings, a two-factor (EPE x ESE) MANOVA was performed on the three major classroom total-scores.

The EPE/ESE taxonomy of subjects provided a means of identifying individuals whose feelings of crowding were correlated with negative reactions to only the physical attributes of their residences, and those who perceptions of crowding were associated with negative reactions to both the physical and social conditions within their residence. The differentiation between these groups was accomplished through a median split of subjects' crowding total-scores, and the subsequent grouping of these scores on the basis of subjects' EPE/ESE response patterns.

It should be noted that the -EPE/+ESE and -EPE/-ESE configurations of data correspond to patterns of neutral and personal crowding, respectively. Thus, by comparing the classroom crowding scores of these two groups through MANOVA procedures, it was possible to assess the relative degree to which neutral and personal crowding experiences in the residential environment affected subjects' sensitivity to crowding in the classroom.

RESULTS

Correlation Analysis

Intercorrelations among the major predictor and criterion variables are presented in Table 1. It is evident that assessments of the residential physical environment were highly correlated with those of physical and social conditions in the classroom. Residential crowding, for example, was significantly related to classroom crowding ($r(31)=.59$, $p <.001$), physical amenity ($r(31)=.45$, $p <.01$), social atmosphere ($r(31)=.47$, $p <.01$), and security ($r(31)=.57$, $p <.001$). The total score for evaluation of the residential physical environment reflected a similar pattern of correlations with the classroom variables. Additionally, size of subjects' residential map (measured by the area covered on an 8-1/2 by 11 inch sheet of paper) was significantly correlated with classroom crowding ($r(31)=.30$, $p <.05$), social atmosphere ($r(31)=.45$, $p <.01$), and security ($r(31)=.41$, $p <.01$).

While the index of residential social atmosphere was not correlated with the classroom total scores, it was significantly related to residential crowding ($r(31)=.39$, $p <.05$) and physical amenity ($r(31)=.53$, $p <.001$). Also, subjects' ratings of people in general were significantly correlated with residential crowding ($r(31)=.42$, $p <.01$), physical amenity ($r(31)=.32$, $p <.05$), and social atmosphere ($r(31)=.50$, $p <.01$). Though alienation from people in general was not correlated with either residential or classroom crowding, an index of perceived alienation from roommates (which was included in the ESE total score) was significantly associated with residential crowding ($r(31)=.57$, $p <.001$). Finally, the summary index of classroom social climate was highly correlated with assessments of classroom physical conditions ($r(31)=.50$, $p <.01$) and security ($r(31)=.56$, $p <.001$).

VARIABLE	R Crowd	EPE	ESE	I-E	Alien.	People	CID	Map Size	Cl. Crowd	Cl. Phys.	Cl. Soc.	Cl. Sec.
Residential Crowding												
EPE	-.82[3]											
ESE	-.39[1]	.53[3]										
I-E	-.08	.17	-.03									
Alienation	.23	-.24	-.36[1]	.28								
People	-.42[2]	.32[1]	.50[2]	-.25	-.42[2]							
CID	-.07	.07	.09	.20	.04	-.27						
Map Size	-.03	.12	.35[1]	-.28	-.34[1]	.23	.17					
Classroom Crowding	.59[3]	-.45[2]	-.03	-.07	-.11	.13	-.05	.30[1]				
Classroom Physical	-.45[2]	.35[1]	.07	.08	.11	-.18	.08	-.17	-.39[1]			
Classroom Social	-.47[2]	.33[1]	-.12	.05	.32[1]	-.28	-.08	-.45[2]	-.25	.50[2]		
Classroom Security	-.57[3]	.50[2]	.19	-.04	-.02	.16	-.19	-.41[2]	-.33[1]	.33[1]	.56[3]	
Course Grd.	.15	-.07	-.15	.14	.36[1]	-.37[1]	.02	-.19	-.07	.18	.35[1]	.40[1]

[1] $p < .05$
[2] $p < .01$
[3] $p < .001$

Table 1. Correlations Among Major Predictor and Criterion Variables (Pearson correlation coefficients provided)

Regression Analyses

In the initial regression analysis, the total scores for residential crowding, physical conditions, and social atmosphere were utilized to predict classroom crowding. All three indices contributed significantly to prediction of the criterion variable ($F(3,27)=5.81$, $p < .01$). The analysis accounted for a total of 39 percent of the variance (see Table 2).

In the second analysis, additional predictor variables were incorporated, namely, residential map size, I-E score, CID score, and alienation from others in general. Again, the indices of residential crowding, physical conditions, and social atmosphere contributed significantly to the prediction of classroom crowding, as did map size, I-E, and CID scores ($F(6.24)=2.88$, $p <.05$). A total of 42 percent of the variance was accounted for in the second analysis (see Table 3).

From Tables 2 and 3, it is evident that residential crowding, alone, accounted for most of the variance (35 percent) in both analyses.

In the third regression analysis, the classroom social atmosphere measure predicted the grades achieved in the course, with residential crowding and the subjects' evaluation of both the physical and social dimensions of their residences improving the prediction significantly ($F(4,26)=3.44$, $p <.05$).

In the final regression analyses, all three residential indices, i.e., crowding, physical conditions, and social atmoshere, contributed significantly to the prediction of visits to the Student Health Center ($F(3,23)=8.07$, $p <.005$) and total visits to health centers, on and off campus, during the academic year ($F(3,23)=5.53$, $p <.01$).[3]

Multivariate Analyses of Variance

In the first analysis, a one-way (residential crowding) MANOVA was performed on the classroom total scores of high- and low-crowding subjects. Results indicated that subjects who felt crowded in their residence rated the classroom environment more negatively than did low-crowding subjects (multivariate $F(4,26)=3.97$, $p <.01$). Univariate analyses revealed that high-crowding subjects felt more crowded ($F(1,29)=9.53$, $p <.004$) and less secure ($F(1,29)=13.53$, $p <.001$) in the classroom than did low-crowding subjects. The former group also rated the physical conditions ($F(1,29=4.63$, $p <.04$) and social climate ($F(1,29)=4.52$, $p <.04$) of the classroom more negatively than did the latter. An additional ANOVA indicated

CRITERION VARIABLE: Perception of Crowding in the Classroom

RESIDENTIAL INDEX	STEP	MULTIPLE r	CUMULATIVE r^2	SIMPLE r*	BETA	RELIABILITY OF REGRESSION
Perceived Crowding	1	.588	.346	.588	.635	$F = 5.81$
Evaluation of the Social Environment	2	.625	.391	-.031	.246	$df = 3, 27$
Evaluation of the Physical Environment	3	.626	.392	-.450	-.060	$p < .01$

* Simple correlation of residential index with criterion variable.

Table 2. Summary Table of Step-wise Multiple Regression: Prediction of Classroom Crowding by Indices of Residential Evaluation.

CRITERION VARIABLE: Perception of Crowding in the Classroom

INDEX	STEP	MULTIPLE r	CUMULATIVE r^2	SIMPLE r	BETA	RELIABILITY OF REGRESSION
Perceived Crowding	1	.588	.346	.588	.619	
Evaluation of the Social Environment	2	.626	.391	-.031	.180	$F = 2.88$
CID Score	3	.634	.402	-.131	-.147	$df = 6,24$
Map Size	4	.645	.416	.149	.142	$p < .05$
Evaluation of the Physical Environment	5	.646	.417	-.450	-.078	
I-E Score	6	.647	.419	-.097	.046	

Table 3. Summary Table of Step-wise Multiple Regression: Prediction of Classroom Crowding by Indices of Residential Evaluation and I-E, Alienation, CID, and Map Scores.

CRITERION VARIABLE: Course Grade

INDEX	STEP	MULTIPLE r	CUMULATIVE r^2	SIMPLE r	BETA	RELIABILITY OF REGRESSION
CL SOC	1	.290	.084	.290	.415	$F = 3.44$
R CROWD	2	.435	.190	.151	.975	$df = 4, 26$
DEPE	3	.579	.336	.067	.756	$p < .05$
DESE	4	.589	.346	-.148	-.101	

(Other variables not entered into the equation were Map Size, CID Score, CL Crowd, CL Phys., and I-E)

Table 4. Summary Table of Step-wise Multiple Regression: Prediction of Course Grade by Indices of Residential and Classroom Evaluation.

CRITERION VARIABLE: Total Visits to Student Health Center During Academic Year

RESIDENTIAL INDEX	STEP	MULTIPLE r	CUMULATIVE r^2	SIMPLE r	BETA	RELIABILITY OF REGRESSION
Perceived Crowding	1	.496	.246	.496	.781	$F = 8.07$
Evaluation of the Social Environment	2	.615	.378	.095	.496	$df = 3, 23$
Evaluation of the Physical Environment	3	.716	.513	-.493	-.534	$p < .005$

Table 5. Summary Table of Step-wise Multiple Regression: Prediction of Visits to Student Health Center During Academic Year by Indices of Residential Evaluation.

CRITERION VARIABLE: Total Visits to Health Centers, On and Off Campus, During Academic Year

RESIDENTIAL INDEX	STEP	MULTIPLE r	CUMULATIVE r^2	SIMPLE r	BETA	RELIABILITY OF REGRESSION
Perceived Crowding	1	.527	.278	.527	.290	$F = 5.53$ $df = 3, 23$ $p < .01$
Evaluation of the Social Environment	2	.580	.337	-.025	.441	
Evaluation of the Physical Environment	3	.647	.419	-.508	-.545	

Table 6. Summary Table of Step-wise Multiple Regression: Prediction of Total Visits to Health Centers, On and Off Campus, During Academic Year by Indices of Residential Evaluation.

that high-crowding subjects rated people in general more negatively than did low-crowding subjects ($F(1,29)=8.34$, $p < .007$).

A two-way (EPE x ESE) MANOVA on the classroom total scored revealed that low-EPE subjects (those who rated the residential physical environment negatively) felt more negative about the classroom environment than did high-EPE subjects (multivariate $F(4,24)=4.23$, $p < .008$). Univariate analyses indicated that low-EPE subjects felt more crowded ($F(1,27)=11.38$, $p < .002$) and less secure ($F(1,27)=14.16$, $p < .001$) than did high-EPE subjects, and also evaluated the physical features ($F(1,27)=4.21$, $p < .05$) and social atmosphere ($F(1,27)=5.99$, $p < .021$) of the classroom more negatively. Neither a significant main effect for ESE nor an EPE x ESE interaction effect was obtained.

In a final analysis, the classroom crowding scores of -EPE/+ESE and -EPE/-ESE subjects within the high residential crowding group were compared to detect possible differences in the generalization of neutral and personal crowding experiences from one situation to another. Results indicated that the means of the two groups were not significantly different.

DISCUSSION

The results of the present study suggest that crowding experiences in residential settings are highly predictive of sensitivity to crowding in at least certain non-residential environments. The correlation and regression analyses, as well as the MANOVAs, provide strong evidence that perceived crowding at home and negative feelings about the residential physical environment are associated with unfavorable reactions to both the physical and social dimensions of non-residential settings.

In support of our first prediction, the regression data revealed that subjects' evaluation of the social atmosphere of their residence contributed significantly to the prediction of their cross-situational sensitivity to crowding. This finding provides some support for a basic assumption underlying the proposed typology of crowding experiences, namely that social factors as well as spatial variables mediate the perception of crowding and the generalization of crowding experiences from one situation to another.

The association between social variables and felt crowding also was reflected in the significant correlations between residential crowding and ratings of people in general, alienation from roommates, and perceived social climate of the residence. In line with these data, a recent experiment by Stokols and Resnick (Note 6) indicated that heightened levels of perceived evaluation and security-threat were associated with increased interpersonal distance and elevated ratings of subjective crowding in a laboratory situation.

The second prediction, that subjects in the personal-crowding residential group would express greater feelings of crowding in the classroom than those in the neutral-crowding group, was not supported by the MANOVA data. The absence of significant differences in classroom crowding between these groups may indicate that personal crowding experiences are not more generalizable between settings than are neutral crowding experiences; or, alternatively, the absence of significant between-groups differences may be related to certain features of the present study. First, small sample size limited the representativeness of our EPE/ESE groupings and the reliability of the related statistical analyses. Second, the fact that ratings of the residential social environment were obtained during the first few weeks of the Fall Quarter may have limited the extent to which social features of the residence could exert a significant impact on subjects' sensitivity to crowding in non-residential settings. In view of the above limitations of this study, the prediction of differential generalization among netural- and personal-crowding experiences remains to be more adequately examined through additional investigations.

The third prediction, that students' visits to physicians throughout the academic year and the quality of their coursework during the Fall Quarter would be significantly associated with ratings of residential crowding, social climate, and physical conditions, was supported by the data. Students' visits to health centers on and off campus as well as their course grade obtained during the Fall Quarter were significantly predicted by subjective ratings of the residential environment, although the quality of course performance was more highly correlated with a measure of social atmosphere in the classroom than with the residential measures.

While the obtained pattern of health and academic performance data generally support our third prediction, it does not provide a basis for inferring causal connections between residential crowding, classroom performance, and student medical complaints. While it is possible that residential crowding experiences promote medical problems and poor academic performance, it is equally plausible that chronically unhealthy or poor students are more susceptible to crowding experiences and social problems than their healthy or academically-gifted counterparts. To provide a straightforward test of our hypotheses, a longitudinal study would be required in which students who are similar in terms of their prior health and academic-achievement patterns could be interviewed over a period of several months with regard to their residential, classroom, and medical experiences. It also would be important to match the students with regard to the architectural features of their residences (e.g., amount and arrangement of space within dormitory suites). A follow-up study which incorporates the above-mentioned design features presently is being conducted at the University of California, Irvine.

The results of this study suggest some additional directions for future research. First, the fact that perceived crowding was more highly associated with social variables within the residential environment than within the classroom setting suggests that the contribution of social factors to crowding experiences may increase as a function of the "primaryness" of the environment. Because subjects spend more time performing personally-important activities in their residences than in a given classroom, the impact of socially-mediated interferences on feelings of crowding would probably be greater in the former setting than in the latter. In order to explore the differential association between feelings of crowding and social dimensions of the environment, it will be necessary to develop a means of locating various environments along the primary/secondary continuum, and to sample the distribution of personal vs. neutral crowding experiences within a diversity of settings.

In developing criteria for the coding of diverse environments, it is important to recognize that the primary-secondary distinction implies both functional and experiential dimensions. Although most persons spend more time engaging in personally-important activities within residential vis-a-vis non-residential settings, individual judgments as to the importance of one's activities and the degree of personal investment in a particular setting are bound to vary. Thus, subjective-report as well as observational criteria should be utilized in attempting to distinguish among primary and secondary environments.

The proposed typology of crowding experiences will be useful to professional designers only to the degree that it predicts specific relationships between environmental dimensions and behavioral patterns. To assess the utility of the typology as a design tool, it therefore will be necessary to examine more thoroughly the linkages between environmental factors, perceived crowding, and corresponding behavioral effects.

Recent experiments indicate that subjective crowding is associated with at least short-term behavioral deficits under certain conditions (cf., Notes 1, 2 and 3; 12). An important task for future research will be to determine more fully the conditions under which crowding experiences result in both immediate and cumulative behavior impairments. The present analysis of crowding offers a theoretical framework for such research.

REFERENCE NOTES

1. Booth, A. Final report: Urban crowding project. Paper presented to the Ministry of State for Urban Affairs, Government of Canada, Toronto, 1975.

2. Dooley, B. Crowding stress: The effects of social density on men with "close" or "far" personal space. Unpublished doctoral dissertation, University of California at Los Angeles, 1974.

3. Evans, G. Physiological and behavioral consequences of crowding. Unpublished dissertation. Department of Psychology, University of Massachusetts, Amherst, 1975.

4. Hanson, L. and Wicker, A. Effects of overmanning on group experience and task performance. Paper presented at Western Psychological Association Convention, Anaheim, California, April, 1973.

5. Keniston, K. Short alienation scales. Unpublished manuscript. Department of Psychiatry, Yale University Medical School, New Haven, Connecticut, 1965.

6. Stokols, D. and Resnick, S. An experimental assessment of neutral and personal crowding experiences. Paper presented at the Annual Conference of the Southeastern Psychological Association, Atlanta, March, 1975.

FOOTNOTES

[1] This assumption is consistent with social learning theory which postulates that one's general expectations concerning the quality of interaction with others will be determined largely by his/her interpersonal experiences in specific situations (cf., 3, 10, 11).

[2] The authors would like to express their appreciation to Ralph Catalano who administered the classroom questionnaires to students enrolled in his Principles of Social Ecology course; and to Susan Miller who assisted in coding the data.

[3] An additional regression analysis was performed on the health center data using only those subjects who occupied architecturally-comparable dorm suites, and had remained in the same suite, throughout the entire academic year (N = 15). This analysis was intended to control for the effects of architectural factors on the relationship between perceived quality of the residential environment and total visits to the Student Health Center during the year. As in the previous analyses, perceptions of residential crowding, physical conditions, and social climate were significantly related to total Student Health Center visits during the academic year ($F(3.11)=3.96$, $p < .05$).

REFERENCES

1. Altman, I. Some perspectives on the study of man-environment phenomena. *Representative Research in Social Psychology*, 1973, *4*:109-126.

2. Desor, J. Toward a psychological theory of crowding. *Journal of Personality and Social Psychology*, 1972, *21*:79-83.

3. Duke, M. and Nowicki, S. A new measure and social-learning model for interpersonal distance. *Journal of Experimental Research in Personality*, 1972, *6*:119-132.

4. Esser, A. A biosocial perspective on crowding. In J. Wohlwill and D. Carson (Eds.) *Environment and the social sciences: Perspectives and applications*, American Psychological Association, 1972.

5. Glass, D. and Singer, J. *Urban stress: Experiments on noise and social stressors*. New York: Academic Press, 1972.

6. Milgram, S. The experience of living in cities. *Science*, 1970, *167*:1461-1468.

7. Mischel, W. Toward a cognitive social learning reconceptualization of personality. *Psychological Review*, 1973, *80*:252-283.

8. Proshansky, H., Ittelson, W. and Rivlin, L. Freedom of choice and behavior in a physical setting. In H. Proshansky, W. Ittelson, and L. Rivlin (Eds.) *Environmental psychology: Man and his physical setting.* New York: Holt, Rinehart and Winston, 1970.

9. Rodin, J. Crowding, perceived choice, and response to controllable and uncontrollable outcomes. *Journal of Experimental Social Psychology,* in press.

10. Rotter, J. Generalized expectancies for internal vs. external control of reinforcement. *Psychology Monographs,* 1966, *80* (whole no. 609).

11. Rotter, J., Chance, J. and Phares, E. *Applications of a social learning theory of personality.* New York: Holt, Rinehart and Winston, 1972.

12. Sherrod, D. Crowding, perceived control and behavioral aftereffects. *Journal of Applied Social Psychology,* 1974, *4*:171-186.

13. Stokols, D. On the distinction between density and crowding: Some implications for future research. *Psychological Review,* 1972, *79*:275-277.

14. Stokols, D. A social-psychological model of human crowding phenomena. *Journal of the American Institute of Planners,* 1972, *38*:72-84.

15. Stokols, D. Toward a psychological theory of alienation. *Psychological Review,* 1975, *82*:26-44.

16. Stokols, D. The experience of crowding in primary and secondary environments. *Environment and Behavior,* 1976, *8*:49-86.

17. Valins, S. and Baum, A. Residential group size, social interaction, and crowding. *Environment and Behavior* 1975, *5*:421-439.

18. Wicker, A. Undermanning theory and research: Implications for the study of psychological and behavior effects of excess populations. *Representation Research in Social Psychology,* 1973, *4*:185-206.

19. Zlutnick, S. and Altman, I. Crowding and human behavior. In J. Wohlwill and D. Carson (Eds.) *Environment and the social sciences: Perspectives and applications.* Washington, D. C.: American Psychological Association, 1972.

THE EFFECT OF CROWDING ON URBAN FAMILIES[1]

Alan Booth

Sociology Department
University of Nebraska
Lincoln, Nebraska 68508

INTRODUCTION

Many, but not all, studies of crowding in animal populations show that compressed living conditions cause marked declines in health, life expectancy, maternal care and increases in aggression and aberrant social behavior (2, 4, 11). While some observers apply these findings directly to human populations, others suggest that man's advanced ability to adapt makes such application questionable at best (6).

Pertinent human research findings are contradictory. Some studies find crowding to be related to decrements in health (8) and in the quality of social relations (9), and to increases in crime (1) and other social problems. Others find crowding to be unrelated to social maladies (5, 3). While recent studies use sophisticated techniques to screen out the effects of such related factors as income, education and ethnicity, with few exceptions their units of analysis are census tracts, communities or some other areal unit rather than individuals and families. Those that do focus on individuals are typically laboratory studies or investigations of institutionalized populations such as students or patients.

The conflicting findings and the unrepresentative units of analysis indicate a clear need for a comprehensive investigation of the effects of crowding at the family level. This is a report on just such a study. The objective guiding the investigation was to

[1] The views expressed here are the author's and not necessarily those of the Ministry of State for Urban Affairs, Ottawa, which provided funds for the study.

discover the effects, if any, of household and neighborhood crowding on such diverse aspects of family life and well being as physical health, mental health, family relations, aggression, relations with kin and neighbors, political behavior, and child development.

RESEARCH PROCEDURES

Characteristics of the Study Area

The thirteen Toronto census tracts included in the study were selected for their potential in yielding a large number of families residing in dwellings in which the number of people exceeded the number of rooms. In some households the number of people exceeded the number of rooms by factors of 2 and 3. The tracts were also selected for the range of compressed neighborhood conditions represented. Population per residential acre averaged from 40 to 150 persons with some individual acres exceeding 600. While not high by world standards, such densities are typical of the majority of cities in North America. The study tracts were older residential areas, although urban renewal and public housing were not uncommon. A range of dwelling unit types (single family, duplexes, row housing, low rise and high rise) are found in the area.

The Sample of Families

The data for this study came from a probability sample of intact, white families of European or North American descent with one or more children, the female of which was under 45 years of age, residing in their present dwelling unit for a period of at least 3 months. From these we selected families so that the number residing in dwellings which had one or more persons per room was nearly equal to the number which had fewer people than rooms. By stratifying the selection of families in this way we were able to control for a number of extraneous factors which might influence the results while at the same time maximize variation in crowded conditions.

Screening interviews yielded 560 families which consented to take part in the study. A comparison of the characteristics of those from whom we obtained interviews with those from whom we did not, revealed that the two groups were similar with respect to occupational status, crowding, age of the head of the household and length of residence in their present dwelling units. The major difference was that many in the latter group had migrated from Western Europe. Husbands and wives were interviewed in depth, separately. They were questioned about family relations and health matters, given several attitude scales, and administered a number of time budget questions. In addition, husbands and wives and all children living in the household were asked to undergo a physical exam at a nearby community health center. More than two-thirds of the families agreed to be examined. To determine whether or not those who agreed to be examined differed from those who did not consent to undergo the physical in some way that would bias our results, we systematically compared the two groups with respect to crowding, health information obtained during the interview, and demographic characteristics. Those who received physical exams had slightly higher socio-economic status, but in all other respects were the same as those who did not. As socio-economic status was controlled in our analysis, we have eliminated it as a source of bias in any findings. Urine and blood specimens were analyzed by an independent laboratory and integrated with data collected earlier by one of the examining physicians. Additional information was obtained on the dwelling units, neighborhood buildings, local commercial facilities, traffic, and other pertinent environmental factors from aerial photos, windshield surveys and public records.

The Crowding Measures

Because research on human crowding is limited and the measures utilized in these studies are quite diverse, no clear standards have emerged for assessing household and neighborhood crowding. For this reason we began our analysis by constructing two simple measures of objective crowding that parallel those which have been used in other studies and examined their effect. The two simple measures we constructed were people per room and households per block face.

People per room has been used in a large number of studies and is constructed by dividing the number of people by the number of rooms (excluding bathrooms and uninhabited rooms such as closets) in their household. Of course, this estimate of crowding does not reflect the actual amount of contact between household members or the extent to which they might interfere with each other's activities. Even though a household may have a high people-to-room ratio, through the judicious management of their schedules, members may avoid (except when sleeping) being at home when the number of people exceeds the available rooms. The measure also contains the built-in assumption that each member (from infant to adult) has the same needs for space.

Households per block face has seldom been used in crowding studies. Dwelling units per acre and dwelling units in the building containing the respondent's household have been used extensively. However, the households in the block face containing the respondent's household plus the households in the opposite block face is a parallel measure reflecting areal congestion. While it has the advantage of representing local conditions, it does not reflect potential sources of congestion such as: automotive traffic, sharing yards and play areas with non-household members, and multiple dwelling construction.

Neither people per room nor households per block face reflects people's perceptions of crowding. Some people may view living in a household with many other people, or in a congested neighborhood, as a pleasant experience, others may find such circumstances frustrating or unpleasant.

The analysis utilizing people per room and households per block face was particularly unpromising. The relationships between pathology and crowding were nil. This factor, coupled with the deficiencies of the two measures outlined above, led us to construct four scales that would more accurately reflect the complex nature of compressed living conditions. We term the scales "objective household," "subjective household," "objective neighborhood" and "subjective neighborhood." The components of each scale were Z scored and summated with each variable receiving a weight of one. By constructing scales we more accurately reflect the many dimensions of crowding while at the same time reduce the possibility of uncovering chance relationships.

Method

The task of determining whether crowding has deleterious (or beneficial) effects and identifying those conditions under which they occur is a complex one. Not only must we screen out the effects of such related factors as income and age, but we must search for catalytic factors such as childhood experiences with crowding. We have screened out the effect of related factors through a statistical procedure called multiple regression analysis. In effect, this procedure permits us to compare people who are crowded with those who are not, while taking into account income, education, and other potential causal factors. It is like comparing low income families who are crowded with low income families who are not, and thereby isolating any effect crowding may have.

FINDINGS

Overview

The effects of the four types of crowding on fifty-eight indicators of stress, poor health, and family and social disorganization were closely examined. In all, 344 relationships were analyzed. In every case, the effects of extraneous factors were screened out. Only 48, or 14 percent of the relationships were statistically significant at the .05 level. A breakdown of the significant relationships

TABLE 1

THE COMPONENT VARIABLES FOR EACH CROWDING DIMENSION

Household Objective

1. Kitchen set in wall of another room.
2. Number of hours respondent is awake and at home when the number of people is equal to or greater than the number of rooms.
3. Number of hours respondent is awake and at home when the number of people in the room with the respondent is two or more.

Household Subjective

1. "Are you troubled by the lack of room or space inside the house here?"
2. "Sometimes people don't invite their friends and relatives into their home as often as they would like because they feel they don't have enough space to entertain them. Do you ever feel that way?"
3. "Home is a place where people get in each others way."
4. "I often feel I don't have enough room to move around in."
5. "Most of the time there are just too many people around."

Neighborhood Objective

1. Type of structure containing respondents' household -- percent in multiple dwelling structure.
2. Number of households in the block containing the respondents' household and on the block face opposite (data was standardized to 600 ft. blocks).
3. Percent sharing space adjacent to the respondents' household (yard, driveway or parking area, sidewalk, courtyard) with non-household members.
4. Street width as an indicator of the amount of automotive traffic adjacent to the household.

Neighborhood Subjective

1. "Are the neighborhood stores in which you shop too crowded?"
2. "What about the places in the neighborhood where your children play, are they too crowded?"
3. "It wouldn't bother me to be able to overhear everyday noises from neighboring homes."
4. "Are you troubled by a lack of room or space where you work?"

indicates that 33 of the 48 pertain to household crowding. All but 3 of these relationships suggest that crowding has adverse consequences. Of the 15 neighborhood crowding relationships, seven indicate compressed conditions have adverse effects and eight show positive consequences. The effects of neighborhood crowding, then, are random and are what we would expect to find by chance. Of the 33 household crowding statistically significant relationships, 12 have to do with objective indicators of crowding and 21 with the inhabitants' perceptions of their conditions.

The remainder of the article focuses on the effects we found. Yet, perhaps the most important finding of this study, contrary to our expectations before we began the study, is that crowded conditions seldom have any consequences, and even when they do the effects are very modest. It is important that the report be read with this perspective in mind, so that the adverse consequences we described in detail do not assume an importance unwarranted by the data.

Health

In our assessment of the effect of crowding on health we chose as indications of health:

1. endocrine responses to stress (blood pressure, urinary protein, urinary catecholamines, eosinopenia, free thyroxine and serum cholesterol);

2. stress diseases such as essential hypertension, angina pectoris and peptic ulcers, and diseases affected by stress such as asthma, upper respiratory infections, neurodermatitis and ulcerative colitis;

3. infectious diseases and communicable diseases;

4. uterine dysfunction;

5. psychiatric impairment as assessed by Langner's 22-Item Index of Psychophysiological Disorder (7). Numerous studies attest to the strength of this instrument as an indicator of mental health;

6. trauma (bruises having accidental as well as agressive origins);

7. people's reports as to whether they stayed in bed because they were not feeling well during the two weeks preceeding the interview; and their estimate of the number of times they consulted a physician during the past year.

Generally, crowded household and neighborhood conditions have little or no effect on peoples' health. Nor are people who feel crowded less healthy than those who do not. Our data suggest, however, that objective and subjective crowding may have a small adverse effect upon health under certain conditions. For example, men seem to be vulnerable to household crowding, especially if they have had no experience with compressed housing conditions as a child. After taking into account the man's age, income, education, ethnic origin, and parents' health, 54 percent of the men who lived in crowded households were diagnosed to have a stress disease, while only 34 percent of those living in uncrowded homes were so diagnosed. The incidence for stress-related disease was 6 percent (uncrowded) and 11 percent (crowded), and for infectious diseases, 17 percent (uncrowded) and 24 percent (crowded).

Women, on the other hand, are unaffected by objective household crowding, but do show evidence of slightly more psychiatric impairment and uterine dysfunction when they "feel" their home is crowded than when they do not. Neighborhood crowding has no effect on health and residence in a congested neighborhood does not intensify the effect of living in a crowded household. Moreover, living in crowded environs combined with the belief that they are compressed has no cumulative effect.

Why men are more adversely affected by objective household crowding than women is not clear. Perhaps early training in maternal care roles facilitates the female's adaptation to the heightened interaction associated with crowding. The fact that men experiencing crowding for the first time are affected most suggests that successful human adaptation to compressed living conditions is quite rapid, perhaps taking no more than one generation. In sum, while we cannot rule out the possibility that crowding has adverse effects on health, our evidence suggests that effects are few and occur only under certain conditions.

Family Relations

Animal studies showing decrements in maternal care to be associated with crowded conditions, and housing studies of human population showing family disorganization to be associated with compressed living conditions have led to suspicion that crowding has adverse effects on the quality of family life. In our study we have attempted to explore a wide spectrum of family ties including those between the married pair and parents and children as well as those between children.

As with health, the findings reported below reflect the data only after the effects of age, income, education, and ethnicity have been taken into account. Only subjective household crowding had an effect on the quality of the relationship of the married pair. Not only did those who felt crowded feel less loved by their spouse, but they also had more arguments with their mate, threatened to leave home more often and withheld sexual intercourse more often than those who did not feel their home was crowded. For example, 51 percent of the men and 42 percent of the women who felt crowded reported their mates were less affectionate, while among those not feeling crowded 38 percent of the men and 30 percent of the women reported love decrements.

Whether or not parents played with their children the preceeding day was unaffected by crowding. However, the number of times parents struck one or more of their children the week preceeding the interview was affected by both objective and subjective household crowding. For example, mothers who lived in crowded dwellings struck their children an average of 7.6 times in the week preceeding the interview while uncrowded ones struck their offspring an average of 5.1 times. The comparable figures for men are 3.7 and 2.9, respectively.

Quarrels between siblings were affected more by the objective than the subjective conditions. Mothers and fathers who were crowded reported more quarrels (9.8 and 8.5, respectively) than those who were not (7.9 and 7.1).

Unlike the medical data, all of the effects of objective and subjective household crowding seem to be intensified by other sources of stress such as economic deprivation. Why the family relations data depart from the medical result is not clear from the analysis.

Social Relations Outside the Family

Studies of the effects of crowding on social relations outside the family have been disparate. Field studies indicated crowded conditions have little effect on visiting patterns and group membership. On the other hand, studies of institutionalized and laboratory populations suggest that crowding leads to withdrawal and decrements in the quality of social relations. We examined the effect of crowding on: visiting friends, neighbors, relatives; joining voluntary associations, attending religious services, job stability; and escapist activities such as excessive sleeping, watching TV, and going on drinking sprees.

Crowded household and neighborhood conditions have little or no effect on people's participation patterns. Nor, with minor exceptions, are people who feel crowded less active socially than those who do not. It is of interest that people who feel that their home is crowded have less contact with relatives in the area. Perhaps relatives place more psychic demands on persons who feel crowded than do friends. That subjective neighborhood crowding is related to voluntary association membership may reflect dissatisfaction with local conditions which are manifested in attempts to alter them through the organization of tenant or other citizen groups.

The above noted exceptions notwithstanding, the fact remains that generally crowded home and neighborhood conditions neither stimulate or retard social participation or signs of withdrawal. Why then should numerous studies produce results to the contrary? Upon inspection, it becomes evident that the studies whose findings are at variance with ours focused on patients and students residing in institutions and on children at play in laboratories. People in such settings may have fewer means available to them to adjust to compressed surroundings. Individuals in crowded households, however, can more readily retreat to another room or even leave their dwellings. Individuals living in a congested neighborhood can find sanctuary in their homes. Neither laboratory subjects, patients nor students have such flexibility in altering their environments as a means of coping with crowded conditions. In sum, while we cannot rule out the possibility that crowding (especially at extreme levels) stimulates or retards social integration, our evidence suggests that it does not.

Political Activity

Crowding has been alleged to increase the incidence with which countries go to war, and to be an important catalyst in civil disorders. While the community we studied had no recent history of either, we felt that crowding might be related to political activities in two ways. First, the psychic stress thought to be stimulated by crowded conditions may lead to aggression, some of which might find an outlet in militant political activity. Second, if crowding is offensive and people attempt to avoid additional social contact, crowded persons may withdraw from traditional forms of political activity that require interaction with others such as joining a political organization, going to meetings and discussing politics with friends and neighbors.

During the course of the interview, the respondents were asked questions about their attitudes towards politics and politicians and about their participation in a wide range of political activities. We examined the effect of crowding both on the approval and participation in politically agressive acts such as demonstrations, civil disobedience and non-approved protests. While none of the crowding measures affected participation in these activities, the number of people in our sample who had actually participated was very small, thus limiting our ability to test the effect of crowding on political aggression. Subjective household crowding, however, was positively related to approval of such activities.

This does not mean that those who approve of protest activities would actually participate if they had an opportunity to do so. It only indicates crowding is a factor which may predispose them to participate or encourage others to do so.

Crowding had absolutely no effect on approving of or participating in traditional political activities such as voting, letter writing, petitioning, election campaign work and so on. However, subjective neighborhood crowding did stimulate neighborhood political activity such as discussing housing problems with friends and neighbors, attending meetings dealing with neighborhood problems and meeting with local officials about neighborhood problems. For example, after taking into account the age, sex, education, income and ethnicity, 34 percent of those who did not feel the neighborhood was crowded took part in a local political activity, while 45 percent of those who felt the neighborhood to be crowded did so. Actual crowded conditions had no effect on these or any other form of political activity. Thus, instead of causing a withdrawal from political activity, crowding has no effect except that subjective neighborhood congestion stimulates local activity.

Reproduction

Studies of crowding in lower animals repeatedly demonstrate that compressed living conditions depresses the ability of the species to reproduce. Crowded conditions have been observed to suppress effective copulation, fertilization, foetal development, infant survival and maternal care. The effect of compressed living conditions has received only cursory examination to date: crowding has simply been correlated with aggregate data on fertility and infant mortality.

From a residential history obtained from each female, as well as the history of each of her pregnancies, and information on contraceptive use, sexual behavior, and family planning decisions, we were able to examine the effect of objective neighborhood and household crowding on: (1) probability of the female becoming pregnant; and (2) probability that any given pregnancy will not result in an infant that survives until one year of age. The probability of foetal and infant survival is of particular interest because among most lower species infant mortality is the primary population control mechanism. If crowding is to influence human reproduction, we thought it should show up in our analysis of infant survival.

Because having a child obviously increases the number of people in the household and therefore crowding, we devised a special indicator of household crowding which removes the people component. We called it the "rooms deficit" measure. Basically it reflects whether or not a family had the same number or fewer rooms than most other families of a similar composition and size. For purposes of comparability we also used the people-per-room indicator. Residence in a multiple dwelling unit was used as an indicator of neighborhood crowding. In our analysis we controlled for the effect of socio-economic status, ethnicity, religious affiliation, years living with present spouse and whether or not the women had been married before.

The crowding measures were not found to influence the probability of the women becoming pregnant nor any of the factors which might influence the likelihood of becoming pregnant such as her general health, contraceptive use effectiveness, or a conscious decision to have a child. Moreover, crowding did not appreciably influence the occurrence of any of the components of infant survival such as induced abortion, spontaneous abortion, stillbirths, or death during the first year after birth. In short, objective crowding was found to have no influence on human reproduction.

Child Health and Development

Surprisingly, little research has focused on the effects of crowding on child health and development. Field studies find crowding to be associated with a loss of parental control, and laboratory investigations show crowding to be associated with social withdrawal. Using the "rooms deficit" measure described above along with objective household and neighborhood measures used throughout the study, we examined the influence of compressed living conditions on health, physical development and school performance.

Crowded household conditions were found to have a small adverse effect on the physical and intellectual development of children. Crowded children are slightly further behind in school than their age peers and are more frequently the object of school authority-parent contacts. Moreover, children living in congested households are shorter, weigh less and are sicker than their uncrowded counterparts. Crowding seems to affect the health and physical development of males, first-born children, children over ten more than females, higher birth order children, and young children. On the other hand, crowding seems to affect the school performance of females to a greater extent than it does that of males. However, it is noteworthy that the effects of crowding are very small. Parental health and socio-economic status, for example, are much more momentous in child health and development than household crowding. Compressed neighborhood conditions were not observed to have adverse effects on the health or development of their young inhabitants.

Why should children be affected by crowded household conditions, but not adults? While our answer is necessarily speculative, we suggest that the adverse influences of crowding may be due to the child's more limited control over his environment and his/her higher physical mobility. The higher status, more powerful adults would have first claim to space in the household and would have a freer hand in manipulating the household environment to their advantage--that is, to reduce the effects of compressed living conditions. Moreover, adults are able to leave the house and the crowded conditions more readily than their offspring. The fact that older and first-born children are affected more than their younger peers lends additional credence to our argument. The spatial needs of adolescents is probably greater than that of adults and yet the opportunities they have to control their environment probably more closely parallels that given to younger children. The resultant stress may account for the decrements in development we observed. The arguments await exploration in subsequent research.

SUMMARY AND DISCUSSION

Conclusion

What may be concluded from the analysis we have undertaken? The evidence suggests that adults adapt to the levels of crowding commonly experienced in Western cities. The fact that men with childhood crowding experience did not suffer the mild decrements in health experienced by those who were encountering household crowding for the first time suggests that adaptation to crowded conditions occurs within one generation at most. Perhaps it would occur in their life times. A study of an older population would be necessary to test this supposition.

The fact that the children were adversely effected by household crowding is consistent with the crowding research conducted by laboratory or institutional settings. Populations with limited control over their physical environments are more sensitive to environmental restrictions, which suggests that there probably are critical levels of household crowding above which it becomes deleterious to health and social relations. Thus, if housing should ever become as constrained as an institutional environment, or as a crowded home is to an adolescent, social and physiological pathology may follow. We believe that the developmental decrements found among the children in our study will disappear once they achieve adulthood and the additional control over the physical environment and reduced physical mobility that goes with it.

The absence of any effect of crowded neighborhood conditions should not be interpreted to mean that the compressed neighborhoods found in today's western cities have no adverse consequences under any conditions. For example, Oscar Newman (10) shows that high dwelling unit density is frequently associated with a design which prevents frequent surveillance of public areas by residents which encourages free access by strangers so as to positively affect crime rates. Because we did not study victimization rates, we cannot say that victimization rates were no different in the crowded and uncrowded neighborhoods.

If our findings are supported by subsequent research, it is clear that humans can tolerate rather high levels of congestion. If the increments in crowded living conditions are not too steep, people can probably accommodate much higher levels of crowding than now exists. Our analysis indicates that higher disposable income would probably eliminate much household crowding. This coupled with the fact that crowding has little or no adverse effect, leads us to recommend that attention and resources should be directed toward income maintenance, health and educational programs, housing design, and other areas that are more directly associated with social pathology. If we do this, the slight effects of crowding on children's health and school performance will either be eliminated or attenuated through better health care and educational opportunities.

REFERENCES

1. Booth, A., Welch, S. and Johnson, D. Crowding and urban crime rates. *Urban Affairs Quarterly*, 1976, *11*(3): 291-308.

2. Calhoun, J. Population density and social pathology. *Scientific American*, 1962, *206*:136-139.

3. Cassell, J. Health consequences of population density and crowding. In *Rapid Population Growth: Consequences and Policy Implications*. Baltimore: Johns Hopkins Press, 1971.

4. Christian, J., Lloyd, J. and Davis, D. The role of endocrines in the self-regulation of mammalian populations. In G. Pincus (ed.) *Recent Progress in Hormone Research, Vol. 21*. New York: Academic Press, 1965.

5. Factor, R. and Waldron, I. Contemporary population densities and human health. *Nature*, 1973, *243*:381-384.

6. Hawley, A. Population density and the city. *Demography*, 1972, *9*:521-529.

7. Langner, T. A twenty-two item screening score of psychiatric symptoms indicating impairment. *Journal of Health and Human Behavior*, 1962, *3*:269-276.

8. Levy, L. and Herzog, A. Effects of population density and crowding on health and social adaptation in the Netherlands. *Journal of Health and Social Behavior*, 1974, *15*:228-240.

9. Mitchell, R. Some social implications of high density housing. *American Sociological Review*, 1971, *36*:18-29.

10. Newman, O. *Defensible Space*. New York: Macmillan, 1972.

11. Southwick, C. and Bland, V. Effect of population density on adrenal glands and reproductive organs of CFW mice. *American Journal of Physiology*, 1959, *197*:111-114.

THE SOCIAL EXPERIENCES OF NEIGHBORHOOD DENSITY AND APARTMENT DENSITY (or CROWDING) IN QUEENS, NEW YORK CITY[1]

Devra Lee Davis

Department of Sociology
Queens College
Flushing, New York 11367

INTRODUCTION

"In the rich literature on the city we look in vain for a theory of urbanism presenting in a systematic fashion the available knowledge concerning the city as a social entity." (31, p. 8). For three very different reasons this statement of Louis Wirth may be as true today as it was when he wrote it in 1938. One likely reason why this is the case reflects the highly political nature of urban studies, on which Robert Merton remarked in 1948 (22, p. 164):

> Yet so deep has been the concern of social movements for housing reform that they have sought to establish a 'case' for adequate housing by citing often defective actuarial inquiries. . ., rather than by directly affirming an institutional right to decent housing in precisely the same sense that education was defined as such a right with the emergence of public education.[2]

The importance of politics notwithstanding, a second, equally important explanation for the dearth of systematic urban theory can be found in an early nineteenth century statement by De Tocqueville which reifies a phobia of size, sui generis.

---

[1] Research funded in part by the National Science Foundation.

[2] Elsewhere in this essay Merton notes that "housing involves the economic interests and social sentiments of important skill-groups and power-groups in American society." The urban researcher encounters hazards from research, institutional cross-fire, urgency and uncritical empiricism (22, pp. 164-177, *passim*).

> I look upon the size of certain American cities, and especially upon the nature of their population, as a real danger which threatens the security of the democratic republics. . . .[3]

Although Max Weber tried to counter this type of thinking in his essay on *Die Stadt*,[4] many other social theorists such as Emile Durkheim[5] and Wirth[6] as well have employed some naive vision of size determinism.

In point of fact, the average population density of American cities has declined by one-half during this century from 6,580 persons per square mile in 1920 to 4,230 persons per square mile. Freedman notes that most major urban centers have been decreasing in population density in the last ten years (16). Yet some recent governmental policies and election results suggest that people

---

[3] Cited by Josiah Strong (29, pp. 128-144).

[4] "Both in terms of what it would include and what it would exclude size along can hardly be sufficient to define the city," *The City*, trans. by Don Martindale and Gertrud Neuwirth. New York: The Free Press (1968) p. 66; also in the very different translation of Max Weber, *Economy and Society*, 3 vols. III, trans. by Guenther Roth and Claus Wittich. New York: Bedminster Press, p. 1213; from Max Weber, *Wirtschaft und Gesellschaft* (30).

[5] Durkheim argued that an increase in the volume and in the density of societies was both the chief cause of the development of the division of labor and the essential element in the development of civilization. "Yet even without recourse to reasoning by analogy, it is not difficult to explain the basic role of this numerical factor. All social life rests on a system of facts emanating from direct and long-standing relationships that have been established among many individuals. Hence social life becomes more intense to the extent that the interactions among the components units themselves become more frequent and dynamic." (11, p. 337).

[6] Although at one point Wirth said that "as long as we identify urbanism with the physical entity of the city. . .we are not apt to arrive at an adequate concept of urbanism," in this same essay he defined the city as a "relatively large, dense, and permanent settlement of socially heterogenous individuals." (31) For a critique of Wirth's acceptance of physical size and density as criteria for the city, *vide*, Emrys Jones (20, p. 8). In his recent book on *Crowding and Behavior*, Jonathan L. Freedman indicates that the cities are not suffering because of their high densities, *per se*. "Although the crime rate in New York 20 years ago was extremely low by today's standards, its density then was the same as it is now, and the same is true of many other cities" (16).

continue to perceive increasing size, sui generis, as an important urban problem. For instance, a number of cities, including Boulder, Colorado, in 1971, Boca Raton, Florida, in 1972 and some Washington, D. C. Suburbs in Maryland and Virginia, in 1970 have passed various restrictions on the number of residential units they will accommodate. The Mayor of San Diego, California was re-elected in 1975 on a campaing of avoiding "Los Angelization." In Forest Hills, Queens, New York, community groups have effectively curtailed several plans for high rise, high density residential construction within the past two years.

This common dread of large size relates to a third explanation for the contemporary relevance of Wirth's criticism of urban theory. While considerable study of urban life has been undertaken, much of it has systematically excluded the social perception of this life.[7] Admittedly the types of information gleaned from aggregate quantitative analyses of demographic shifts differ substantively and epistemologically from those based on open-ended interviews. And ideally both types of information should be included in any study of urban life. Much research, however, has been based exclusively on hard, aggregate quantitative analyses, with little attention to the role of peoples' perceptions of their urban lives. And where peoples' perceptions have been found "inaccurate," as in the Booth project (3, 4), their significance has been discounted. Thus, Booth finds that the feeling of being crowded is not highly related to actual conditions. Some feel crowded who are not by external standards and others, living in very compressed conditions, often do not feel crowded.

What such studies ignore is the overwhelming importance of social perceptions. Social perceptions are not less critical for being objectively inaccurate; in fact, they may be all the more pertinent on this count. Esser (13), and Hall (19) concur on the significance of cultural mediations of space and living conditions. Karlin, Epstein and Aiello (12) point out a related problem of many studies correlating population characteristics and measures of social pathology: such studies focus on the empirical antecedents and consequences of population characteristics, while failing to consider the human coping processes.

As Merton noted some time ago in his famed elaboration of the Thomas Theorem, whether or not a given situation can be empirically confirmed, if significant numbers of people believe it to be true,

[7] Jerome Bruner and Leo Postman, give this delimitation of social perception, "What/ a person/ sees in any situation and how/ that person/ sees it inevitably reflect the manner in which he approaches the situation, the way in which he is *eingestellt*." (5, p. 71).

this, in itself, has real sociological consequences in explaining that situation. Thus, in Merton's study of the social organization of Craftown (22), it was noted that a large proportion of residents were affiliated with more civic, political and other voluntary organizations than had been the case in their previous places of residence. It was further noted that this was also the case for those residents having infants and young children, a somewhat unusual finding, given the difficulties for parents in lower economic levels to secure the child care necessary to pursue such social activities. The Craftown parents explained their involvement as due to the great number of teenagers available for taking care of younger children. However, this explanation from the everyday world of Craftown parents did not mesh with the empirical analysis of Craftown's demography: the ratio of adolescents to children under ten years of age in Craftown was 1:10; compared to a ratio of about 1:1.5 in communities lived in before coming to Craftown (23).

In Craftown the critical factor appears to have been public perception: people perceived that there were more teenagers available to work as babysitters, therefore they consequently behaved such that they utilized teenagers as babysitters more so in Craftown than in their previous residences.[8]

METHODOLOGY

Taking into account the preceding issues, the present project explores the effects of apartment and neighborhood density on the quality of urban life focussing particularly on the social perception of density.

Following Merton's lead and parallelling some of Booth's recent methodological directives, an analysis was conducted of residents' social perceptions of neighborhood and apartment density as compared with the formally legitimated descriptions of these two components of urban density.

[8] "The extent to which that behavior in turn reinforced the initial perception is, in this case, secondary. Craftown is less an instance of the self-fulfilling prophecy than of public perception and consequent behavior," Robert K. Merton, personal communication, 1976.

TABLE I: RESEARCH DESIGN

	Neighborhood densities	
	high	low
Social structure variables and intra-dwelling density constant	Rego Park	Little Neck

	Intra-dwelling densities	
	high	low
Social structure variables and neighborhood density constant	Jamaica I	Jamaica II

Two areas of identical socio-economic status, neighborhood density and ethnicity, but differing intra-dwelling (or home) densities were compared, in order to study the significance of intra-dwelling (or home) density. Two areas of comparable socio-economic status, intra-dwelling densities and ethnicity, but differing neighborhood densities were compared, in order to consider the significance of neighborhood density. Apartment density was measured in terms of the number of people divided by the number of rooms (minus the kitchen and bathroom); and neighborhood density was measured by the ratio of people to the unit area. As has been noted by other critics, both of these density measures do not reflect the actual amount of contact between neighborhood residents or between household members; still, they are used here as adequate, typical representations of apartment and neighborhood density.

To give some illustration of the typical high and low neighborhood densities, consider that the average floor area ratios, units per acre, and units per block, respectively, were: 2.9 and .23; 139 and 9; 1,104 and 30. The typical high and low density apartments had ratios of people to rooms respectively of .5 and below, and 1.0 and above. The quantitative analysis of these areas is discussed further in Davis, Bergin and Mazin (7), and Davis (8, 9).

The geographical areas of interest here were selected because of their quantitative differences in dwelling and neighborhood densities. However, much of the analysis of these areas entails qualitative, unsystematic, naturalistic observation with its attendant restrictions.[9] Given its concern with the social perception of density, this study necessarily dwells on activities generated by face-to-face interactions--an arena that Goffman has aptly termed "the field of public life" (18, p. ix). An approach in the ethological genre has limitations and advantages. This posture is not affected as a rejection of all hard, empirical observation, but as a recognition of the value of corroborative or supplementary analyses beyond the pale of replication. With Goffman, it is assumed "that if a broad attempt is to be made to tie together bits and pieces of contemporary social life in exploratory analysis, then a great number of assertions must be made without solid quantitative evidence" (18, p. xiv). To round out the picture, some attempts at more systematic, quantitative study are included here as well.

Time limitation in our study and an interest in employing independent measures of the same phenomena stimulated the development of a paradigm using three simple methods for examining peoples' experiences of their apartments and their neighborhoods: in-depth interviews, shorter questionnaires, and unobtrusive measures.

In-depth Interviews

A snowball technique was utilized to obtain respondents, keeping several demographic quotas in mind. Respondents were questioned in unstructured three hour interviews about the quality of life in their neighborhood and the adequacy of personal living space in their apartments. People were asked what they liked best and least about each; in what ways they found their area changing; what their living plans were; how and where their children played and studied. For these interviews residents of the respective neighborhoods, but not of the specific blocks, were employed.

---

[9] Goffman (18, p. xvi) comments on the corresponding hazards of more traditional research designs: "In spite of disclaimers, the findings of these studies are assumed to hold more broadly than the particularities of their execution can immediately warrant; in each case a second study would be necessary to determine of whom and what the results are true. The variables which emerge tend to be creatures of research designs that have no existence outside the room in which the apparatus and subjects are located, except perhaps briefly when a replication or a 'continuity' is performed under sympathetic auspices and a full moon. . . . Frameworks have not been extablished into which a continuously larger number of facts can be placed. Understanding of ordinary behavior has not accumulated; distance has."

Shorter Questionnaires

A second technique involved the administration of close ended questionnaires, providing some background data about neighbors and the neighborhood. People were asked: how many friends do you have who live in your building (or within 200 yards); how many people in your building (or within 200 yards) do you know; how many can you name; how many do you normally say hello to; do you lend or borrow things from your neighbors.

Unobtrusive Measures

Unobtrusive measures of peoples' responses to strangers at their doors were also included as part of our study, in order to determine whether people responded differently to strangers depending on the characteristics of their apartment or their neighborhood. These findings are discussed in Davis (8, 9). Other unobtrusive techniques were employed, including photographic and twenty-four hour sound recordings, which provided supplementary pictures and sounds of the respective neighborhoods. To study typical neighborly exchanges, researchers noted the proportion of people who would respond when asked for the time, for directions, or for change for a bus. In all cases, interviewers consisted of people who lived in the neighborhood.

ISSUES

With the preceding techniques, this study explores the following issues about urban density: What is the relationship, if any, between observed neighborhood street activity, and apartment density (9). Following Stokols in viewing density as a necessary antecedent (26) rather than a sufficient condition for the stressful experience of crowding, how do residents perceive crowding in their primary (home or main) environments as compared with crowding in their secondary (or side) environments. Goffman (17) considers that a main involvement will typically be a dominating one and a side involvement a subordinate one, implying that disruptions in the former would be less tolerated than in the latter. Stokols has hypothesized (26) that social interferences arising from conditions of high density or proximity would be potentially more disruptive and frustrating in primary than in secondary environments. One indirect assessment of these ideas can be obtained from in-depth interviews and from unobtrusively noting the responsiveness of residents to peoples' requests for help in their neighborhood streets (secondary or side environment) and at their apartment front door (main or primary environment). A higher rate of response to our interviewers on the street than at the front door would constitute support for the preceding, insofar as it indicates that people respond more frequently to intrusions into their secondary or side environments.

Relationships between residents is another important topic for this study. Numerous social analysts have taken for granted that the intensity of neighborly contacts can be graded in an order reported by Alderson (1) and Keller (21): (1) villages; (2) old established working class areas; (3) new estates; (4) residential quarters in own house property. Baldassare (2) considered that urban crowding, sui generis, causes individuals to have less friends, see their friends less often, and know their neighbors less intimately than their counterparts in less dense environments. The comparison of the observed and reported social interactions of residents in Little Neck and Rego Park, provides evidence of the effects of neighborhood density on neighboring. Where people have less semi-public neighborhood space, as in Rego Park, they may be expected to report and evidence less neighborliness.

To study the independent effect of apartment densities on neighboring, two groups of residents from more and less dense apartments in Jamaica are examined. Additionally the effect of ethnicity and class on neighbor interactions can be inferred from this comparison. If Rego Park and Jamaica II exhibit similar kinds of neighborly activities, this would suggest that neighborhood density, sui generis, is an overwhelmingly important factor as Baldassare (2) and others imply.

And if these two areas display different kinds of neighborly activities, this would indicate the importance of class and ethnic factors, as Carey (6) and others have noted. This latter finding would add further support to Carey's argument that different ethnic populations manifest different crowding thresholds.

Activities of children in all the households and in the neighborhoods presents an interesting issue, especially in relating peoples' reports of these with their unobtrusive measures. Here, we hypothesized that people would report vast differences in the types and places for play, reflecting class and ethnicity, and lesser but important differences reflecting apartment and neighborhood density. In Jamaica I and II we expected that children would play more in the neighborhood than in the apartments and that children in the more dense apartments would be restricted in their home activities. In both Rego Park and Little Neck, we expected a wider range of play activities, but we expected that children in Rego Park would be subject to more supervision than those of Little Neck.

In both the reported and observed studies of neighborly activities and those of children's behavior, this project was structured so as to gather evidence about class and ethnic influences. The considerably more detailed work of Booth (3, 4) reports little significant correlation of such factors with urban density; however, his work says nothing about non-white ethnic influences, as the entire sample consisted of white Canadian/Toronto families of European or North American descent. The lack of consistent findings on the effects of density and crowding reported in numerous other studies doubtlessly reflects problems of research design; and studies like Booth's do much to overcome some of these problems. But findings such as Booth's should not be construed as evidence that the effects of crowding are universally few.[10]

RESULTS[11]

Little Neck

Of 89 questionnaires completed in Little Neck, 54 were obtained from female respondents, and 35 from males. Little Neck is predominately white, Jewish and middle class. The sample interviewed for this study largely reflected the census distributions of such variables as age, ethnicity, education, family size and length of residence. As the Appendix shows, the ages of those responding to our questionnaire in Little Neck are lower than those in Rego Park, reflecting the greater number of teenagers in Little Neck.

As was done for Rego Park, a question concerning the length of residence at the current address was asked, as a measure of neighborhood stability. On the average, people in Little Neck have lived in their current houses for approximatley 11 years. This stability is in sharp contrast to Rego Park, where residents tended to stay in their apartments only until they accumulated enough money to purchase a house.

---

[10] In point of fact, although Booth's work considers only white families, some of his findings do support the contention that there are significant ethnic and class variations in responses. Hence, more families of western European descent declined to participate in the interviews, while more families of higher socio-economic status opted to participate in the physical examinations. In a similar vein, the work of Schmitt (24, 25) in Hong Kong and Honolulu does not demonstrate that high densities are universally unrelated with urban stress; rather it indicates the socio-cultural mediations of such phenomena.

[11] See also Appendix I.

The Little Neck area is composed of singular, detached one- and two-story houses. Each house has a plot of land approximately 45 by 110 feet in area, which is generally well-maintained. This area has been designated sound by the New York City Planning Commission. Built as a development, the houses are in rows about 30 feet from the streets. Each has a driveway and some have garages in which to park one car. Most homes are strikingly-similar in architecture both outside and in, as possible models of Tom Lehrer's ticky-tacky-houses-all-in-a-row. They are usually one-story frame houses consisting of six rooms and a basement. A scattering of houses are larger--two-stories and eight rooms. Many of those interviewed had finished their basements to provide additional rooms.

The streets running through the area are tree-lined. Little traffic is noticed whether it is during the day or evening, as none of the streets is a through-street. Only three or four outlets are available onto Little Neck or Grand Central Parkways. Thus, most traffic is generated from inside the area by residents. While heavy traffic is not a problem on these streets, parking vehicles is problematic, since many families own more than one car. In some instances parking leads to disagreements, particularly in the area around the local community pool which borders on two residential streets and Little Neck Parkway.

Other community facilities besides the pool are the local YMHA and a nearby schoolyard. Children in the neighborhood utilize all three areas for play in general, although the pool and Y are on a membership basis. The younger children can be seen playing in the yards around the houses for the most part; an observation supported by our interview data from the residents. Older children generally play stickball and other games on the uncrowded streets or in the local schoolyard and parks, where teenagers can be found congregating in groups in the evening when the pool is closed. Trouble with teenagers is not reported by most of the interviewed residents. Neighborhood facilities are thought to be generally safe by the respondents. As in Rego Park, a wide variety of stores is not available.

Most of those interviewed drive to a small shopping center composed of a supermarket and a handful of other shops, located a few blocks away north of the bordering Long Island Expressway.

Those interviewed stated that people rarely socialize in the streets or on the sidewalks of the area. In fact, during the summer months few kids play in the streets and around the houses, since most use parks, playgrounds, or the pool area. Adults are generally inside their air-conditioned homes during the daytime, and may sit outside with one another after dinner in the evenings.

Respondents generally were divided on whether the schools are too crowded. Often they gave a qualified "yes" or "no"; depending on the particular school. The grade schools are thought to be underutilized while the junior and senior high schools are considered to be overutilized.

The typical community family is nuclear, consisting of four members, usually two children. Questions were put to those Little Neck residents, seeking the specific things liked and disliked about their own residences. With most homes having three bedrooms, each child has his or her own room in addition to the spacious family rooms. Overcrowding is reported to be virtually nonexistent. The typical Little Neck home (in 46 out of 50 cases) has six or more rooms. Most of the time (in 39 out of 50 cases) there are four or less living in these homes. Eighty-five percent of those interviewed stated that they liked the fact that their homes are spacious. Living in a convenient location between two controlled access thoroughfares, on a good block, and in a good neighborhood, are specifically listed desirable qualities. Following these on the list of things liked about the house is the privacy made available by living in a "private" house. Residents mentioned having their own room, their own yard, a family room or finished basement, and more than one bathroom as definite plusses for their houses.

As for complaints with their residences over one-third of the respondents mentioned "nothing" that they disliked about their house. Some want more land and space between houses, while only 15 percent complained of inadequate facilities *inside* of the house. These dislikes were concerned with poor layouts, small rooms, lack of sufficient closet space and thin walls.

Specific things liked and disliked about the neighborhood were elicited also. Friendly, good neighbors are overwhelmingly among the most appreciated aspects about the neighborhood. Also appreciated are the uncrowded and shaded streets, picturesque scenery, and quiet. Residents noted that it is a good place to raise kids, and that it is a stable, middle-class area.

The largest number of people expressed no dissatisfactions with their neighborhood. A small minority indicated dislikes about Little Neck, such as poor transportation within Queens. Transportation to and from Manhattan is excellent and inexpensive for the Little Neck resident. For transit within Queens one needs a car. Another dislike about the neighborhood was that the houses were too close together, and parking two or three cars is difficult. Many complained that their homes have boring, monotonous exteriors.

According to both reports and observations, interaction among neighbors seems to be well-developed in the Little Neck community. During many of the interviews which were conducted in front of the houses, as neighbors passed by greetings were exchanged. Often, some neighbors were together at each other's homes. Many times after completing the questionnaire with a resident, he/she would tell you to try the next-door neighbor or would inform you that they were not at home at that particular time.

Upon investigation, it was found that most people have a number of close friends living in the neighborhood. In fact, many purchased their homes in the area because friends of theirs had moved in and were extremely satisfied. Reasons for such a high degree of interaction stem not only from the stability of the neighborhood, but also from the nature of private-house living. The property in front of the single-house provides a meeting ground for neighbors and friends.

As can be easily predicted, residents on the block almost invariably know more than ten of their neighbors. This relationship is not superficial or casual. Most adults are of the same age groups and have children that are growing up together. Also, the relationship existing between immediate next-door neighbors is much more well-developed than it is in Park City. Little Neck respondents said that they know both their next-door neighbors well enough to help each other out at least. More significant than that is the large number who visit frequently with each other, be it alone or with their spouse.

The following are some excerpts or reports from our in-depth interviews with residents of Little Neck.

One female resident for ten years spoke of the kindness of her neighbors who cared for her children while her parents were hospitalized. She also mentioned communal neighborly activities such as joint ventures for shopping, attending films, making barbecues and birthday parties.

One male resident for seven years reported the development of a neighborhood alarm system, whereby when one hears a neighbor's burlar alarm or sees anything suspicious in the area, she or he immediately notifies the police. This man also told of a good deal of borrowing and lending of tools, describing most of his neighbors as "pleasant, helpful people who will take the time to lend a hand."

Another resident said that "living is much easier here than in the city. There is always a place to go when one wants to be away from everyone else. There is also the chance to sit in my own backyard which is private but out of doors."

One father reported that "children always have enough room to study. Of course they bring a lot of friends in with them and I don't know how much real schoolwork they do, because we have six television sets[12] and they are always watching. But you can be sure that they do not fight over which programs to watch."

Many of the residents said that walking in the neighborhood in the evening was no problem: "It's good to be able to run out and visit in the evening without calling to tell family that I arrived all right," one teenager said. She mentioned, "we have very little crime here, even now. I joined the Rape Task Force Project and then it turned out that we had almost no work to do in our own neighborhood, because that just doesn't seem to happen here very much." She added, "Besides, I feel that if I really had any hassles on the street, people here would help me out. This isn't Manhattan, you know."

A young mother of three especially liked being able to send her children outside to play. "I don't worry about having them outside, even though my youngest is two. Of course, I have to watch the cars, but I don't have to watch out for strange people."

One resident complained about being in contest with his neighbors. "I was trying to decide what kind of second car to buy and it hit me that I had better get something a little snazzy, sort of to go with the neighborhood." When asked what "'snazzy'" meant, he replied, "you know, not a second-hand car, but a new one, or at least one that looked new. Everybody knows everybody's cars around here."

Another resident said "I wished our houses didn't look all alike. But I guess that inside we make them different enough with drapes and whatnot. We finished our basement and made it a den; then my kids took it over and now it is an unofficial clubhouse for the junior hippies here. It's cute and soundproof (thank God), so the blaring music doesn't bother us. We just had to get an extra phone for all the calls the kids started getting there; but they pay for it out of their babysitting money."

The unobtrusive measures employed in Little Neck included twenty four hour block observations, which revealed some amusing aspects of neighborhood life in midsummer. When we first set up our taping equipment to record variations in sounds, we left this equipment in a light paper bag on the grass and waited in a parked car nereby. From the car we had hoped to record our own observations and also to watch over the recording equipment during the night. However, people walking on the street stopped, poked at the bag, and invariably upset our recording which then had numerous scratching sounds and occasional sonic yells of "Hey, John, you know whose this is?" or "Whose kids are fooling around with this here?" At one point several people were gathered around trying to decide whether the

---

[12] This was an exceptional number of television sets; most families did own two televisions.

recording equipment had been stolen by some peripatetic thief or dropped by some forgetful neighbor. For related reasons, it was difficult for our 'unobtrusive' 'nonparticipant' observer to retain here status, while she sat in the car during the night. Ultimately, she confided to a resident that she was doing a research project for school. Whenever any other resident approached her, which happened up to midnight, the informed resident (who watched from his stoop) waved off the approacher, signaling with a hand motion that everything was all right.

The responses to our door-to-door canvassing in Little Neck are reported elsewhere (8, 9). A similarly high rate of cooperation was elicited when residents were asked for the time, for directions or for change. Out of 100 requests for each of these assists, no one was turned down flat (or not responded to) when asked the time or when asked directions: the minimal reply noted was "I don't know," or "no idea." The request for change did elicit a ten percent non-response, but this has an obvious explanation, given many urbanites' fears about mugging. More interesting about these exchanges was the tendency of people to ask if they could be more helpful, or to offer some additional comments, such as: "Are you late?" "My watch may not be correct." "Have you ever been there before?" These comments extended assistance to the strangers.

Rego Park

A total of 72 questionnaires were completed in Park City. Of these, 43 respondents were female, and 27 were male. Those interviewed were predominantly white and Jewish and middle class, reflecting the census distributions for the area, although the neighborhood itself is becoming less homogenous.

The modal length of residence at the current address was no more than five years. The second most frequently reported length of residence was less than one year. At the other end of the scale was a small group, who have been tenants in the development since its opening 17 years ago.

Park City is composed of six high-density high-rise buildings located on a superblock in Rego Park, Queens. In addition to the apartment buildings, a membership pool and a rental garage are provided for those who wish to pay. Management rules and three-feet-high hedges prohibit use of the immediately surrounding, well-maintained lawns for any reason. Concrete paths channel people from doorways to streets. Hence residents must leave the property in order to congregate outdoors. Surrounding the Park City property are four heavily-trafficked streets of which three abound with stores, affording many types of shopping. A bordering street provides an asphalt playground and a series of other similar buildings called Park City Estates, owned by the same corporation as Park City. Only two blocks away is the Long Island Expressway and Lefrak City, the largest apartment complex in Queens.

The streets around the circumference of Rego Park are extremely crowded during the daytime and evenings. The numerous shops and stores draw people not only from the immediate area, but from all areas of the borough. Inside the stores, long lines are often evident. Those interviewed generally feel that most people are in the streets during the afternoon hours and both a.m. and p.m. rush hours. In addition to the regular shopping crowds, many children are on the street after three p.m. when school is released. Two-thirds of those responding to our questionnaire stated that car traffic on their block was too heavy.

Children in the neighborhood play freely in many areas of the dense neighborhood. Where they play generally tends to be related to the age of the child. For example, the youngest kids are in the much used playground across the street playing in the sandbox, or on the swings. As they become older, they can be viewed playing ball in the schoolyard separated from the playground by a high fence. The teenagers tend to congregate around the benches on the periphery.

Residents' responses to the questions on recreational facilities appear to be about equally divided as to whether they are adequate or not. Most qualified their answers with regard to the age groups. For the youngest, they stated facilities were adequate; for the older kids, they said "no."

The park is owned by the City and during the daytime hours a park attendant is in charge. This fact, plus the fact that the park is crowded with many residents, gives many residents a sense that the children are safe.

Most Park City apartments are not near the overcrowded point of 1.0 or more persons per room. Children generally have a room of their own or share a room with another sibling of the same sex. Some 70 percent of those reponding stated that the apartment was large enough for their family's needs.

Other questions sought to elicit statements that would delineate the specific things liked and disliked about the apartments. The following were things liked by Park City respondents about their apartments. The thing liked most is the terrace, which all apartments in Park City have. The terrace effectively becomes the defensible, semi-private "backyard" for Park City residents. A second listing was the largeness and/or spaciousness of the rooms. Many of those interviewed felt that they were as large as one could find in any similar middle-income development. Closely following room size in response is the location or convenience of the apartment. Many list this as the main reason why they chose to live in this development. Also cited to a great extent was the bright and airy quality of the apartment. Because these apartments do not have air-conditioning, an airy apartment is highly regarded by these residents, especially in the summer. A number listed the layout as being one of the things they liked about the apartment. Following closely was the kitchen, with its modern appliances.

Things disliked about the apartment: Significant here is the fact that the first listing is "nothing" or a blank in place of another form of response. The second most frequent reponse is maintenance and problems related to sanitation. Dislikes such as these do not occur in any particular apartment, but rather are attributable directly to the number of people sharing the same incinerator or elevators, etc. Also disliked about the apartment by respondents is the size of the rooms which they consider to be too small. It is interesting to note that comparable numbers of people thought their apartments to be too small as found them large enough. It may very well be that such a disparity exists not only with subjective understandings as to what is "big enough," but also with the size and amount of furnishings in the apartment. The last significant characteristic listed that is disliked is a lack of modern conveniences such as air-conditioning, dishwashers, and other kitchen appliances.

Specific things liked and disliked about the neighborhood were elicited also. Convenience to a wide range of shopping is clearly high on the list. Some 60 percent of those responding cited the stores as a definite asset to the neighborhood. In addition to this, over half listed convenience to transit. (Manhattan is a 20 minute, one-token ride away.) Following the above, a clean, quiet, and a relatively safe neighborhood was included in the plusses for the area.

As previously mentioned, this area appears to be burdened by traffic and congestion. Parking problems and related traffic problems are cited as some of the major things disliked about the neighborhood. Also, dirty streets and related sanitation problems are objected to. Many papers and dogs' exretement line the sidewalks and curbs, and make the area unsightly. Sanitation is hindered by the traffic and parked cars around Park City.

Neighborhood social interaction between residents of Park City is limited. Few neighbors exchanged "hellos" or recognized one another either on the streets or in the elevators. Residents tend to keep to themselves except, in some instances, possibly to chat in the park with those of the same age group. The typical Park City resident knows his immediate neighbors only on a "Hello, how are you?" basis, if that at all. Only ten percent said that they were friendly with their neighbors even to the extent of sometimes visiting them. Often residents do not know the neighbors on their floor.

As in Little Neck, a number of questions of this matter were included in our questionnaire. For instance, some 60 percent of those interviewed stated that their best friends reside outside of the neighborhood. This is significant because it appears that residents do not go out of their way to make new friends when families move in. Ten percent of the respondents said that their best friends live in the building. Forty percent of those interviewed know less than 10 people in their building even well enough to say hello to. Those who do know many neighbors well enough to say hello emphasize the point that they acknowledge them in the elevator or on the street, but hold almost no serious conversation with most of their neighbors.

In order to explore this further, a social interaction scale measured to what degree a respondent interacts with her immediate next-door neighbors. For both neighbors, the largest majority stated that they talked to them occasionally. This accounts for 50 percent of all those interviewed in Park City. Approximately 20 percent know their next-door neighbors only by sight or well enough to say hello to. From these data, 70 percent of Park City residents engage in only casual interaction with their next-door neighbors.

Some 20 percent do know their next-door neighbors well enough to borrow things or to help each other out. Approximately 5 percent fall on each extreme--they wouldn't know them if they saw them on the street; and they visit frequently with one another.

Excerpts from our in-depth interviews further elaborate the experiences of Rego Park residents:

An elderly woman said: "In my old neighborhood on Martense Street, people would talk to you. Each night they would congregate in someone else's house. The neighbors would ask you if you wanted something from the store. The people here, when I see them in the elevator, look at me as if I were a wild animal. You could die here and no one would care."

An elderly widower reported: "The neighbors that used to live next door to me were extremely noisy. They were always fighting. I didn't complain. They left at the point that I was about to break. . .Oh Christ! The walls are so thin. You can sneeze and the guy next door to you says 'God bless you'. . .I am not friendly with my neighbors. The people are very cold-hearted. On the elevator, no one says 'Good morning.' When I first moved into the building,

I tried to say hello to people but they didn't answer back and after a while I stopped trying."

An adolescent male resident for 6 months said: "I don't speak to my neighbors too often. I don't like them. If my neighbors ever complained about me, I'd shut them up. Once I went out of my way to hold the door for my next-door neighbor and she didn't acknowledge it. Next time, I'll slam the door in her face."

One 35 year old woman resident for 5 years said: "The elevator in this building is like a foreign country. It seems like the people from Hungary all fell off the boat into the elevator. Ninety percent of the people here don't understand you. Do you think that in this building you know who your next door neighbors are?"

One person said: "I know my neighbors well." When asked what does this mean, she responded: "I say hello to them. I wouldn't borrow things from them, though."

A mother of a 6-month-old child reported: "The people here are terrible. You could die in the hallways and no one would help you. I have never seen people like this. They don't care about anyone or anything. I don't talk to my neighbors other than for a few bangs on the wall or them yelling at me to turn my stereo down. . . I know only three people here."

Some did complain though that they didn't get as much privacy as they would like. This complaint was most often voiced by young mothers with children. In the words of a 29-year-old housewife with children ages 4 and 7, "I threaten to put locks on the door--to provide a place where the kids wouldn't be allowed, but my friends here warned me about making a room off-limits. I am never really alone during the day and as a consequence I stay up late at night." In another case the attrition of family solidarity resulted in a "twin" separation:

"There are now four people living in this apartment. My mother, my 19-year-old sister, my 13-month-old baby and myself (23 years old). Previously there were six people living here. The family has just experienced a twin separation. My husband and father have just left. We have three bedrooms, two bathrooms, a terrace, kitchen and a dining room. For two families, there just wasn't enough room. The apartment was meant for one family. We moved here when my husband came out of the service. The lack of space was a definite factor in our hassles. . . . Sometimes it got to be so crowded especially if friends were being entertained in the living room. There isn't a circular type of seating. The bedroom is the only place where you can get peace and quiet."

It was possible to complete the unobtrusive sound recording in Park City. Measured in this fashion the streets were quite noisy, contrary to the residents' reports. However, the noises recorded consisted mostly of non-human sounds, such as cars bumping and parking and stopping. Very little human conversations were picked up and almost no shouting or greeting exclamations. One conversation recorded was of two males disputing who had claim to a parking space. Our paper-bagged recording equipment remained unbothered through the twenty-four hour period, as did our non-participant observer resting in his car.

The observer confirmed the absence of people on the streets, except for apparently limited activities such as entering or leaving the building, parking cars, transporting groceries and delivering or picking up packages. Double-parking and street congestion were high, particularly between four and seven in the evening.

In the attempt to conduct our measure of cooperation by asking residents for the time, directions or for change, we encountered some special problems unique to the neighborhood. Since relatively fewer persons were on the streets at any given time, it took three times as long in Park City as in Little Neck to find one hundred persons to ask. And, when we did this, out results corroborated the Shadow Scale for this area. (8, Table 3). Most often our inquirers were not able to complete their request for assistance, so that effectively the residents of Park City refused to help without knowing what they were being asked to do. We used all male researchers for this task. Although data were not gathered systematically, we later used two female interviewers and found that they received more assistance in Park City than their male counterparts; but even the females elicted very limited kinds of help in this area, when compared with that that the males received in Little Neck.

As would be expected based solely on the architecture of Park City problems arose in the daily life in that could not occur in a different living structure. For instance, the elevators were usually quite crowded; and often they are delayed by maintenance people collecting garbage or by tenants moving in and out. On the elevators themselves residents are normally silent. Although the riders in the elevators are in fact most often neighbors, there is no interaction between them, reflecting the fact that these riders are strangers to one another.

The laundry room was quite busy, having 10 machines for 210 families. It would be even more busy, if a significant number of people did not report sending their laundry out, for fear of using the rooms. For those who do their laundry in the building, some strategies have developed to serve their own interests in completing the task quickly, given the limited availability of machines. People displayed behavior here much like that in conserving parking spaces: they staked their claims. In the laundry room a common strategy is to place coins in the machines, while a person is preparing the clothes. This stakes the machine quickly. Of course, it also makes others wait, until they self-defensively learn to adopt the same strategy.

Jamaica

The Jamaica neighborhood was considered in two subsamples: more and less dense apartments. The completed questionnaires numbered 70 less dense and 80 more dense. Distributions in regard to ethnicity, age, sex and length of residency appear to be similar for both predominantly black, lower class groups. Of the 70 uncrowded questionnaires, 50 respondents were females, and 20 were completed by males. Similarly, 57 females and 23 males completed the questionnaires making up the overcrowded group. The age distributions of the respondents are in the Appendix.

In regard to the stability of the project, the average length of residence for the less dense group is 8.5 years. Closely following is the more dense segment which has an average length of residence of 7.5 years. Both groups contain many families who have lived in the project since it was built, either 18 or 32 years ago depending on the particular section. Although not as stable as Little Neck, Jamaica is a great deal more stable than Park City. This can be attributed partly to the better quality housing in the project when compared to what is available for this group in the private sector.

The Jamaica Houses I and II are situated in the South Jamaica section of Queens in New York City, which has been designated "deteriorating" by the New York City Planning Commission. The numerous buildings of the project comprise a super-block some two blocks long and two wide. Inside the project are mostly three-story brick and concrete structures with equally shared numbers of seven-story elevated and five-story walk-up structures. The three- and five-story structures are placed side-by-side in long rows across from one another so that a courtyard is formed between them. Small lawns are planted near the buildings; the rest of this area is devoted to the concrete courtyards. Around the older buildings of Jamaica I, trees provide ample shading for the benches surrounding the courts. The Jamaica II buildings generally are lined along the streets and take into account the seven-story structures. Most of the front entrances are right on the sidewalks which carry all the auto traffic and the bulk of the pedestrian traffic surrounding the project. Noise levels in these buildings (and especially the apartments fronting the street) are much higher than those in apartments facing the rear and surrounding courtyards. Land-use in the Jamaica Project appears to be much better planned overall than that of Park City. There is ample public room to sit on benches and congregate in groups. Also, smaller children can play in the courtyards in full view of their parents from the apartment windows. Small mini-playgrounds for toddlers are provided in numerous sections of the project.

The buildings themselves all follow the same design. Regardless of their size, they are all of concrete and brick construction. None have terraces. The lobbies are small, devoid of sitting room; they serve only as entrances and mailbox areas. All the hallways in the building are tiled in a dark beige color. The stairwells of the smaller buildings smell of animal urine, and have many children playing in them. According to many of the children, they are directed to play there by their parents who insist on keeping them close by. Very frequently, it is reported that children do play on the sidewalks and streets, or "anywhere they can."

Surrounding the project are a number of different environments. To the west are the Long Island Railroad tracks, which are elevated from the ground on top of an embankment. The passing trains are easily audible inside the apartments. The few streets that provide access to the other side of the tracks occur at the periphery of the project. To the south and east of the project are located numerous one- and two-family structures. Many are abandoned and boarded up, with the rest being "deteriorated." Often these are claimed by junkies. Approximately two blocks eastward from the projects is New York Boulevard which is described by the New York City Planning Commission as an area that, with the exception of its density, is comparable to the ghettoes of Harlem, Bedford-Stuyvesant, and the South Bronx. To the north of the project are numerous locations for both heavy and light industry. Most notably are the junkyards with metal materials stacked-up in huge piles (not yet redefined as urban art). Abandoned well-stripped autos and abandoned turn-of-the-century houses pepper the areas between industry. With the sole exception of the South Jamaica projects, the entire neighborhood is "deteriorated."

Not surprisingly, most of those interviewed when asked to list things they liked about the neighborhood either did not respond or stated "nothing." Those who answered differently conceded that Jamaica was merely "OK" or "nice." These respondents could not elaborate on exactly *what* was "OK" or "nice" about the area. Significantly, a large number reported satisfaction with neighbors. Of course, this says nothing about the physical "deteriorated" characteristics of the area, but provides evidence that neighbors can obtain emotion and psychological support in battered neighborhoods.

In contrast, asking the respondents to delineate things they disliked about the neighborhood yielded a wide variety of answers. The most frequent response was that there are too many junkies on the streets. Others cited the related but less controversial topic of "lack of safety" and "high crime" as things disliked about the neighborhood. Other often listed phenomena were "everything," run-down condition, vandalism, and no stores in which to shop.

The South Jamaica Project appears to have a high level of social interaction. Based on observations, most residents know one another in their buildings at least on a casual, greeting basis. Standing in the lobby by the mailboxes, one can hear neighbors talking casually among themselves while picking up their mail. Often, in front of the buildings or in the courtyards, older and younger residents congregate in small groups and pass time together. While riding in the elevators, neighbors would exchange "hellos" quite frequently. We asked residents where their best friends lived. Over half stated that their best friends resided either in the neighborhood or in the building.

For the less dense group, over one-half of the respondents know 10 or more people in the building at least well enough to say hello to. If the group who answered that they knew 6 to 10 people is included, then 70 percent of the total responses are accounted for. Of the 30 percent who know between 0 and 5 people, the majority were either elderly or new residents or an employed person. For the overcrowded group, a great deal more respondents knew 10 plus people in their buildings. In fact, two-thirds answered in this category.

Interaction among next-door neighbors was investigated to obtain a clearer picture of the depth of these relationships. The most common reports were that each knows the other well enough to "help each other out" or to "visit with frequently." Similar to the other measures of interaction, and probably for similar reasons, the neighbors of overcrowded apartments cited these statements as the best description of their relationship with their next-door neighbor more frequently than the uncrowded group. Excepting these two categories, the other most often cited relationship was that they "talk together occasionally." Extremely few respondents reported that they "don't know their neighbors even by sight" or that they know them "only by sight."

The streets bordering on the South Jamaica I and II Projects are busy. Many children can be seen playing in front of the doorways and on the sidewalks. Older kids in their teens often are found talking or walking in small groups on the sidewalks. Additionally, many women stand or sit outside conversing or walking with neighbors and friends. Although the sidewalks in South Jamaica are bustling, they are considerably less crowded than those of the Park City area. The

South Jamaica Project area has only four or five small stores that generate a small volume of shoppers, in contrast to the hundreds of varying shops that line Park City's streets. Pedestrian traffic in South Jamaica, then, consists of neighborhood residents involved in social interaction as opposed to consumerism. Not only is the quantity of traffic in Jamaica diminished, but the very quality of pedestrian traffic is altered. South Jamaica sidewalks are paced by more leisurely and communicative groups whose very reason for being on the sidewalk is social. (of course, many pedestrians are performing tasks generally not social; i.e., shopping, visiting a friend's apartment, etc. But the great majority, based on observation, consider the sidewalks public space in which to interact with friends.) People can be observed on the streets at almost any time of the day or night. In fact, residents report in equal amounts that most people are on the streets in the morning, afternoon, evening, late at night, and at "all times." Apparently, then, crowding on the streets is not in and of itself a problem, but is rather a part of the natural order of events of this community which has little internally defensible space.

Because most of the families in the project do not own automobiles, parking on the street was not a "traffic" problem (as it was in Little Neck and especially Park City) although speeding and racing cars were. Residents often reported that before our project began in the South Jamaica area, two children were killed by speeding autos in separate incidents. Residents had to fight for a number of years in order to obtain a traffic light on one corner. According to the residents, crowding is also a factor in the local schools. Only a few respondents stated that the schools were not over-utilized.

Many of the apartments appeared to be furnished well and comfortable to sit in. Others, however, were cluttered with children's playthings, clothes, or furniture. Some residents complained about peeling paint on the walls or ceilings due to leaks from the apartments above.

One of the most significant differences between less dense and more dense Jamaica occurred in regard to the social experience of crowding. The more people in a given size apartment, the more that living space is taxed. In the less dense apartments, only some 25 percent interviewed stated that they lacked space or that the apartment was not big enough for their needs. In contrast, 60 percent of those respondents residing in more dense apartments cited space and crowding problems.

As expected, crowding problems appear more dramatically in things liked and disliked about the respondent's apartment for the more dense group. This, too, proved to be the case. Residents in the denser apartments stated twice as often as those in the less dense units that there was not enough room. In the same proportion they also cited poor maintenance as a dislike. In a unit overutilizing the facilities of the bathroom, kitchen, etc., breakdowns occur more often. Delays incurred in repairs effect greater numbers of people in the overcrowded units.

Both Jamaica groups listed some similar things liked and disliked about their apartments. Respondents considered the amount of rent charged, the ease in ability to keep the apartment clean, and the particular floor on which they lived, as definite assets for the apartment. Significant dislikes were the high noise levels audible in the apartment, and the lack of screens or window guards.

On the question of whether there was enough space to entertain guests, both groups generally stated in proportionate numbers that there was "always" or "sometimes" enough space to entertain. However, more than twice as many people in the more dense group said that there was "never" enough space to entertain guests. The incidence of children not having a place to do homework and not being able to bring friends into the homes was listed *three times* as frequently in the more dense apartments. This same difference can again be noted for how often one has the opportunity to be alone when she wants to be. In the more dense apartments, responses indicate that one "never" has the opportunity to be alone when she wants *three times* more often than the respondent in a less dense unit.

Differentiating between the less and more dense apartmental living conditions within the high population density low-income project in South Jamaica, it was discovered that in the denser apartments, the respondents reported discomforts that were unknown to those living in apartments in the same neighborhood that were not crowded. Listed below are highlights of selected in-depth interviews in the denser units:

(1) A family of 12 (6 boys, 4 girls, 2 parents) living in a 7-room apartment (5 bedrooms). There isn't enough room for the family; and they cannot afford to move. In this environment, it is impossible for the children to study. The children aren't allowed to bring friends into the house because the mother feels that "I have enough of my own to look after." They can't eat together because of the lack of space. There is barely enough room to sleep, let alone for personal privacy.

(2) Seventeen people living in a 5-bedroom apartment. No friends of children are allowed in the house. "Got enough of my own in here. I have problems telling which one of them's mine. There's always noise; always a mess. We can't hear the neighbor's noise over ours."

(3) "There are seven people in my apartment, with five rooms. The rooms do not serve my family's needs. They are too small and there aren't enough of them. . . We have a very small refrigerator which is very run down. The kitchen itself is very small. . . The three boys and two girls sleep together in the same room. . . Since we live in such confined quarters, it is impossible to live decently. There is no privacy of any kind. . . The kids do not study. . . All five children share two beds and one dresser. There is no place to be alone. . ."

"When the neighbors get noisy, we bang on the walls. The neighbors are very noisy. This is very bothersome and leads to arguments. Even the neighbors argue amongst themselves. Most of the people keep to themselves. We know very few people in the building. Since we are just trying to keep our heads above water, we have little to no time for making new acquaintances. Most people keep to themselves and never help others even in emergencies."

Conditions of poverty also prevailed for project dwellers who lived in less dense apartments, which had more rooms than people. They, too, were on welfare and those who did work were toiling at low prestige jobs. Nevertheless, they tended to be more satisfied with their existence as compared with residents living in crowded apartments. They tended to argue less amongst themselves, get along better with their neighbors, several of whom they could name; also they had a higher appraisal of the neighborhood; and their children were better students. For example:

(1) There are 3 living in a 4-room apartment (formerly there were 4). The one that moved out has graduated college and is going to law school this year. The 18-year-old is hoping to follow in his footsteps.

(2) Two retired people living in a 3-room apartment love their neighbors. They sit outside on the stoop and converse with them, chronicaling the neighborhood's activities.

(3) Four people iving in a 6-room apartment. Occasionally, there is some shouting in the morning but not much. There isn't that much competition for the bathroom and kitchen facilities. The children enjoy the privacy of their own rooms and are able to study. The parents welcome the children's friends at any time. The residents especially the children, spend a good deal more time inside the house and less time outside.

(4) "We have five in six-and-a-half rooms. The kids have plenty of room. . . Our neighbors visit us often and we get along really well with them, exchanging floor and favors."

Of course, outside of the confines of the apartment, both the crowded and the non-crowded people have to combat the squalor of the "black" ghetto. The facilities, ranging from the subways to the supermarkets to the schools are hopelessly overcrowded. Many would like to move out, but their miserable economic situation prevents them from escaping. Almost everyone questioned admitted to being scared of opening the door to strangers; yet, in fact, doors were opened frequently to our interviewers in the Shadow Scale study.

In conducting the unobtrusive sound recording in Jamaica, the noise levels remained consistently high throughout the twenty-four-hour period, save for six hours from 12:30 a.m. to 6:30 a.m., when the levels were considerably more quiet. With the considerable litter on the busy streets our bagged, disguised tape recorder was not disturbed once as it lay on the sidewalk. Automobile sounds constituted one section of the noises, but human exchanges were equally predominant, reflecting the highly social nature of the street. The observer confirmed the busy street scene and was twice approached, while parked in his car: once he was asked what he was looking for (a possible entree to selling or entrapping him for illegal drugs); and the other time he was asked by a police patrol car to state his business. Children were observed selling parking spaces to those residents seeking them; and they also engaged in playing chicken with moving cars, positioning themselves so as to force a car to stop suddenly.

In measuring neighborhood cooperation by asking direction, etc., a very high rate of response was obtained, black interviewers receiving somewhat less assistance than white interviewers. As opposed to Park City's elevators of strangers, the elevators of Jamaica were places of conversation and greeting. Often jokes might be made about the "new dude riding up" (the observer); and speculation was offered about whatever anyone would want in Jamaica: "tell the brother, this only goes right back down, and he looks like he goin sideways." At certain hours the laundry rooms in Jamaica were full of people, some doing laundry, some socializing and tending to children. Based on the observation of numerous clothes lines, ingeniously hung from high rise windows, the laundry rooms are not used by all the tenants for laundry although they are definite congregating places.

ANALYSIS

Given the qualitative nature of the data reported here, no attempt can be made to provide statistically significant observations about the effects of neighborhood and apartment density on the quality of urban life. In fact, this study must be viewed not in opposition to but as a supplement to the types of information included in formal statistical analyses.

With these provisos in mind, certain characterizations of the areas studied can be made. Middle income residents of the high density Rego Park neighborhood report more social problems with their neighborhood and neighbors than do their counterparts in the low density neighborhood of Little Neck, even though unit density of both areas is low. Typical remarks of residents of the crowded neighborhood of Rego Park were that the neighbors were unfriendly, the area was not safe and that parking was a problem. Only forty percent knew ten people to whom they rountinely said hello. During interviews and the indirect requests for assistance, residents of Rego Park rarely greeted one another or offered assistance. Their responses to strangers at their apartment doors were abrupt, incomplete and often nasty. Residents of the less dense neighborhood of Little Neck generally showed satisfaction with both the neighborhood and their neighbors.

The observed differences in street exchanges with strangers imply corresponding differences in the region of space on public streets in which mutual presence prevails. What Goffman has termed the full conditions of copresence (17, p. 124) may well vary with the density of the neighborhood: in a neighborhood of higher density, people tune out many of those they encounter on the street. Literally they do not mutually recognize one another even as potential residents of the same area. To attribute these differences primarily to the respective neighborhood densities of Rego Park and Little Neck would be folly; because there are economic and ethnic factors that bear here as well, as other observations indicate. "One might say, as a general rule, that acquainted persons in a social situation require a reason not to enter into a face engagement with each other, while unacquainted persons require a reason to do so." (17, p. 124) The responses to requests for assistance on the street met with much greater success in the less dense neighborhood of Little Neck as compared with denser Rego Park and it also met with great success in both sections of Jamaica. Obviously the 'reasons' for face interaction are not simply neighborhood-density dependent. Still, for white, middle class Americans in Rego Park and Little Neck, these findings do allow speculation that living in a perceived crowded neighborhood makes people more resistent to strangers, or less likely to engage others in face encounters. And it is conceivable that echoes of political solidarity among lower class Black Americans analogously sustains their sense of mutual presence, regardless of their neighborhood density. It is not that density is irrelevant, but that it must be viewed in a particular sociohistorical context.

In Jamaica residents reported considerable problems with the neighborhood, regardless of their apartment density. However, residents of less dense apartments complained more about the neighborhood, while those of more dense apartments complained more about their neighbors. Typical remarks of those living in denser apartments were that they had no privacy; children cannot study or bring friends into the home; and guests are not welcome because there is not sufficient space. Residents of these apartments reported that their neighbors were unfriendly, noisy, and rude. However, when unobtrusive measures were used, all residents of Jamaica displayed cooperative, helpful behavior, both to strangers at their door with the Shadow Scale and to the various requests for assistance.

This discrepancy between the way people in Jamaica behave, which is on the whole helpful to one another, and the way they say they behave is an interesting phenomenon. One possible explanation is that residents of Jamaica do not consider their everyday exchanges with neighbors to be as friendly and open as they might be under better circumstances. They have an idea of neighborliness which involves different relations than they now have. Another possibility is that what they report about neighbors reflects what they feel they should say to any interviewer. And a related explanation is that the complaints about neighbors may refer mostly to immediate next-apartment neighbors, about whom they have frequent, reasonable complaints, given the poor sound insulation and other design problems.

In part, this can be taken as support for the hypothesized greater intensity of social interferences in primary (main) environments as opposed to those in (side) secondary ones. People complain most of frustrations and disruptions in their apartments than they complain about those in their neighborhood. Certainly the unobtrusive studies of assistance on the streets support this, insofar as people were far more willing to grant assistance in semi-public places (i.e., secondary environments) than they were at their front doors (i.e., primary environments). As Stokols has suggested (27), stress responses are greatest where private, personal space is most directly infringed upon. In this respect, the experiences of residents in higher density apartments in the high density neighborhood of Jamaica involved more direct personal thwartings, while those of residents in lower density apartments in the high density neighborhood of Rego Park involved mostly neutral thwartings.

While this project did not extensively differentiate friends from neighbors, it did gather interesting information on different neighboring relationships in the areas studied. Contrary to some of the ideas of Keller (21) and consistent with some of the findings of Baldassare (2), residents of denser neighborhoods did not exhibit less neighboring when measured in terms of the number of friends and kinds of friendships than those of more dense neighborhoods. However, here class and ethnicity play important roles. The black, lower class dense community of Jamaica has a distinct character from that of the Jewish, middle class, dense community of Rego Park. The suspicion that both class and ethnicity are important here is based on reports from other neighborhoods.

No simple relationship between neighboring and neighborhood density appears from the reports here. Certainly the marked differences between the high density neighborhoods of Jamaica and Rego Park attest to the importance of class and ethnic factors. Although both these areas are high density, unobtrusive studies of neighboring revealed that residents of Jamaica showed considerably more neighborly helpfulness than did their counterparts in Rego Park; yet residents of both areas directly reported problems with their neighbors. In addition to the explanations presented earlier for this discrepancy, it is possible that residents of Jamaica directly report the common sense view of neighboring in a high density neighborhood; they perceive themselves to be less neighborly simply because they have so many neighbors. As Baldessare (2) notes they may in fact have as many intimate relations with their neighbors as do residents in less dense neighborhoods, but this is a decidedly smaller proportion of the total neighborhood.

In the study of children's play, perhaps the most outstanding finding is the range of possibilities. Children in Little Neck have their own backyards and common frontyards where they play at very early ages (sometimes pre-walking) with minimal supervision. Children in Rego Park and their parents are restricted to the immediate apartment until they are old enough to attend supervised day-care programs, which often tend to be indoors. Children of both sections of Jamaica are also restricted to the immediate apartment until they are considered to be old enough to go out with their older siblings. Sometimes they are enrolled in day-care facilities, but these are subject to the vagaries of state, city and federal funding and are in no way easily available to all who could use them.

Children up to pre-teenage are watched over by their parents in Rego Park, much more so than in Little Neck or in Jamaica. In the latter case this often seems to reflect a confidence that the children must develop their own street savvy. Some parents do not share this attitude and occasionally a mother in a less dense apartment who does not work outside the home will order her children to come straight home from school and not to go out without permission. In such cases a great deal of domestic conflict is not uncommon, as after school the streets (in summer) and the hallways (in winter) are normally flooded with youngsters running, playing and skipping rope. Of course, parents of children in denser apartments complain about having no space for children to play inside their apartments, but they are powerless to change this. Obviously the lack of supervision of the pre-teenagers in Little Neck reflects different circumstances. There are organized after-school activities, including swimming, arts and crafts classes and sports. Children in Little Neck tend to congregate in one another's houses where they often have special playrooms or dens.

DISCUSSION

From the observations reported here and from numerous recent studies discussed in this volume and elsewhere (16), there is no necessary relationship between urban density and the quality of urban life. Thus, while higher urban densities may be thought to account for some problems of urban living, in almost no cases is density itself responsible for these problems.

One difficulty for any theory of urbanism clearly relates to problems of delimitation and operationalization of urban density and urban crowding. Until a few years ago, social theorists implicitly equated the two, often taking any situation of great density to be ipso facto a situation of great crowding. This sort of size determinism was soundly criticized on several fronts as part of the general interest in the quality of life and in cultural mediations of physical circumstances. Unfortunately, some work which attempted to rectify this previous methodological oversight, tended to overcorrect the problem in a peculiar fashion. If the past dilemma had been the failure to distinguish crowding and density, the new solution was to make this distinction, study respective correlations or regressions, and then conclude that this was clear evidence of the specific effects of crowding or density. Lost in this process was the importance of intervening variables, which could themselves account for many of these findings.

For instance, Schmitt's findings in Honolulu and Hong Kong (24, 25), Draper's observations of the !Kung Bushmen (10),[13] and other studies of non-western cultures all indicate the significance of cultural factors for the social experience of high densities. Even within modern western culture, itself, however, no simple generalizations can be drawn about the effects of high urban densities, independent of sub-cultural and economic factors. Thus, Booth's project on Toronto families which shows minimal effects of apartment or neighborhood density (3, 4), Stokols et al's small-group studies of student dormitory living (28) which demonstrates different experiences with primary personal thwarting may represent findings specific to the populations and the living situations that they studied.

Considerable physiological evidence indicates that whatever its etiology the *feeling* of being crowded results in an arousal state in the brain (14). This feeling can occur under conditions of high or low densities of living (15). Given the importance of this feeling

---

[13] Although their population density is approximately one person per 10 square miles, the !Kung Bushmen live in intentionally compact arrangements so that conversations between different family circles are overheard.

for the development of physical responses, careful attention to its sociological sources can now be considered. We might well augment the Merton-Thomas theorem: When people believe a situation to be crowded it is real in its physiological consequences, as well as in its sociological consequences.

APPENDIX I: CHARACTERISTICS OF NEIGHBORHOODS STUDIED[14]

	Little Neck	Rego Park	Jamaica I	Jamaica II
SAMPLE SIZE (Questionnaire)	89	72	80	70
0-9 age range	17	3	7	12
10-19 age range	13	5	6	9
20-29 age range	14	24	15	19
30-39 age range	10	16	11	28
40-49 age range	11	5	11	15
50-59 age range	15	10	9	2
60-69 age range	1	3	4	0
70+ age range	2	2	8	0
ETHNICITY	black working class	black working class	predominantly white, Jewish middle class	
AVERAGE LENGTH OF RESIDENCE	11 yrs.	2 yrs	7.5 yrs.	8.5 yrs.
NEIGHBORHOOD DENSITY	low	high	high	high
DWELLING DENSITY	low	low	high	low
TYPICAL DWELLING	single-family home, 1 or 2 stories	high-rise apartment	high-rise apartment	high-rise apartment

[14] These are general characterizations, not to be confused with survey estimates.

REFERENCES

1. Alderson, S. *Britain in the Sixties: Housing.* Harmondsworth: Penguin, 1962.

2. Baldassare, M. The effects of density on social behavior and attitudes. *American Behavioral Scientist,* 1975, *18*:815-825.

3. Booth, A. Urban Crowding Project: Final Report. Ottawa: Ministry of State for Urban Affairs, Canada, 1975.

4. Booth, A. The Effect of Crowding on Urban Families. This volume.

5. Bruner, J. and Postman, L. An approach to social perception. In *Current Trends in Social Psychology.* Pittsburgh: University of Pittsburgh Press, 1948.

6. Carey, W. Density, crowding, stress in the ghetto. *American Behavioral Scientist,* 1972, *15*:495-508.

7. Davis, D. L., Bergin, K. J. and Mazin, G. When the Neighbors Get Noisy We Bang on the Walls: A Critique of Density and Crowding. Paper presented to the American Sociological Association meetings, 1974, expanded version in *Man-Environment Systems* (forthcoming).

8. Davis, D. L. The shadow scale: An unobtrusive measure of door-to-door interviewing. *Sociological Review,* 1975, *23*:143-150.

9. Davis, D. L. Density, crowding and stress: A theoretical model and an unobtrusive measure. In *Systems Thinking and the Quality of Life.* Compiled by Karl C. Blong, Washington, D. C.: Society for General Systems Research, 1975.

10. Draper, P. Crowding among hunter-gatherers: the !Kung Bushmen. *Science,* 1973, *182*:301-303.

11. Durkheim, E. *Division of Labor in Society.* George Simpson (trans.) Glencoe: Free Press, 1947.

12. Epstein Y., Karlin, R. and Aiello, J. Strategies for the Investigation of Crowding. This volume.

13. Esser, A. H. (ed.) *Behavior and Environment: The Use of Space by Animals and Man*. New York: Plenum Press, 1971.

14. Esser, A. H. A biosocial perspective on crowding. In J. Wohlwill and D. Carson (eds.) *Environment and the Social Sciences: Perspectives and Applications*. Washington, D. C.: American Psychological Association, 1972.

15. Esser, A. H. Toward a definition of crowding. Guest Editorial in *The Sciences*, October, 1971.

16. Freedman, J. *Crowding and Behavior*. New York: Viking Press, 1975.

17. Goffman, E. *Behavior in Public Places*. New York: The Free Press of Glencoe, 1963.

18. Goffman, E. *Relations in Public, Microstudies of the Public Order*. New York: Basic Books, 1971.

19. Hall, E. T. *The Handbook of Proxemic Research*. Washington, D. C.: Society for the Anthropology of Visual Communication, 1974.

20. Jones, E. *Towns and Cities*. London: Oxford University Press, 1966.

21. Keller, S. *The Urban Neighborhood*. New York: Random House, 1968.

22. Merton, R. Social psychology of housing. In Dennis Wayne (ed.) *Current Trends in Social Psychology*. Pittsburgh: University of Pittsburgh Press, 1948.

23. Merton, R. *On Theoretical Sociology*. New York: The Free Press, 1972.

24. Schmitt, R. C. Implications of density in Hong Kong. *Journal of the American Institute of Planners*, 1963, *29*:210-217.

25. Schmitt, R. C. Density, health and social disorganization. *Journal of the American Institute of Planners*, 1966. *32*:38-40.

26. Stokols, D. On the distinction between density and crowding: Some implications for future research. *Psychological Review*, 1974, *79*:275-277.

27. Stokols, D. The experience of crowding in primary and secondary environments. *Environment and Behavior*, 1976, *8*:49-86.

28. Stokols, D. Ohlig, W., Resnick, S. M. Perception of residential crowding, classroom experiences, and student health. This volume.

29. Strong, J. *Our Country: Its Possible Future and Its Present Crisis*. New York: Baker & Taylor, 1885.

30. Weber, M. *Grundriss der verstehenden Soziologie*. 4th ed., Winckelmann (ed.). Tubingen: J. C. B. Mohr, 1956.

31. Wirth, L. Urbanism as a way of life. *American Journal of Sociology*, 1938, *44*:1-24.

SECTION 3: MICRO-ECOLOGY

INTRODUCTORY NOTES

The three papers in this section each treat individual and small-group behavior primarily within the context of small-scale built environments. As in earlier chapters, the common theme is the manner in which people regulate interactions with other people who may be known to them or whom they wish to keep from being known in a particular situation. Two of the papers also attempt to place such behavior in a larger theoretical framework.

Zimring, Evans and Zube provide an overview of research and refer to many of the concepts addressed in Sections 1 and 2. They examine the literature, not only of *crowding*, but also of the corrolary concepts of *personal space* and *small group ecology*, treated more or less as sub-fields. Noting the continuing conflict between the holistic outlook of the designer and the analytic methods of the laboratory scientist, they call for a systems approach which can combine these. They offer a multi-variate model, which is not deterministic, that is, it allows for the fact that in complex real world settings there is rarely a single cause for behavior or a single consequence of a design decision.

Zimring, Evans and Zube introduce three relatively new research concepts. The first of these is the degree of perceived *control* that people have over their environment and the effect it has on their behavior and satisfaction with designed space. The second, *cognitive mapping*, is related to the first in that it comprises the selective experiences stored in the mind which allow people to make sense out of the environment in order to deal with it. The third is *arousal*, a phenomenon long known to neurologists but only recently considered by behavioral scientists. Arousal denotes the state in which the nervous system is "turned on," to be ready for various environmental stimuli and appropriate responses. Its relationship to crowding was discussed in the previous section. But as a state of readiness, it relates to the control and mapping concepts in that it pertains to the individual's propensity to seek or avoid stimulation. One type of stimulation which appears to govern arousal is interaction with other people. In arousal, we have both a possible cause and effect of regulating the boundary between privacy and communality. Although the authors caution that the nature of the connection between arousal and particular environmental elements has yet to be demonstrated clearly, the concept offers the promise of assessing responses without relying on verbal reports which are always problematical. If it turns out to be an accurate means to measure environmental responses directly, it may become extremely useful in design evaluation.

DeLong focuses on context, the relationship between interpersonal transactions and the places in which they occur. This will be of particular interest to designers, for whom context is everything. DeLong makes a distinction between *contextual logic* and the more traditional linear logic based on *a priori* values. In contextual logic, it is *relationships* that matter; discrete events or "facts" have no meaning except as they relate to others. Behaviors in one environmental context become functionally different in another, even though by traditional logic they are "same." Function is central to that basic design concept, *structure*. Structure is defined here as the limitations on the freedom of occurence of events, in other words, of behaviors. This in turn is the source of *stability*, but in a paradoxical way because stability requires constantly changing input in an unvarying context. Thus, DeLong addresses that fundamental relationship of all art, of all design, of life itself: unity in diversity. Constancy, or stability, can only be achieved in the midst of variability, and vice versa. Privacy and communality are complementary and opposite unities--variability of one increases stability of the other. As he puts it, "It is the distribution, or allocation, of variability which leads to perceptual and/or functional invariance." Allocating variability is, indeed, a useful definition for design. In DeLong's view, it is *relationships*, rather than isolated events or objects, that are the functional units which designers must use to introduce stability into any environmental system, and they can do so only when they stand in contrast to one another, in other words, through variety. As such, DeLong's argument relates to both Esser's and Greenbie's presentations in the beginning and the closing chapters of this volume.

This paper is an unusual blend of abstract theory and vivid example. Illustrations range from experiments with the effect of furniture arrangement on conversational groups to the conversations of children, visual illusions and mathematical relationships. These are developed in two conceptual contexts: Hall's theory of the use of space as a communication system (proxemics) (3) and Altman's view of privacy as a boundary regulation process (1).

The final paper in this section, by Efran and Baran, reports on an empirical study which illustrates quite well a number of the points raised in the two previous papers. Through the use of a pressure sensitive switching mechanism in a floor mat, in conjunction with deliberate violation of male toilet norms by an experimenter, the behavior in a men's room was examined to explore the differing privacy needs at the urinals and the washbasins. The counter under the mat preserved the anonymity and, therefore, the privacy, of the subjects, but the experimenter of course did not, since the object was to invade privacy. (Appropriately enough, the men were psychologists at a conference!) This experiment vividly supports DeLong's thesis; privacy in the context of urinating was quite different from that in the context of handwashing. It also demonstrates quite well his point, which is also developed in several other chapters in this volume, that privacy as boundary regulation in Altman's terms can be effected through environmental factors (stall partitions and doors) as well as verbal and postural cues. The common point here is the "cut-off," visual and otherwise, which Coss hypothesized as a design principle (2). It also may be suspected that arousal will vary considerably from one context to the other for the same basic cues in these different contexts, and devices to measure arousal might usefully elaborate such findings. Efran and Baran make the important point (consistently stressed by Hall) that nonverbal communication often takes place out of awareness, while conceptualized boundary controls are often based on verbal symbols. They also make the significant observation that, while privacy needs are usually thought of in terms of relatively long time-intervals, actually urban life enforces numerous short-term interactions, where privacy needs, nevertheless, are extremely important to the welfare of the individual and may influence social behavior in significant ways. They also illustrate the psychological and physical consequences of design decisions which fail to provide for these needs, in this case, toilet doors.

REFERENCES

1. Altman, I. Privacy: A conceptual analysis. *Environment and Behavior*, 1976, *8:* 7-29.

2. Coss, R. G. The cut-off hypothesis: Its relevance to the design of public places. *Man-Environment Systems*, 1973, *3:* 417-440.

3. Hall, E. T. *The Hidden Dimension*. Garden City, N. Y.: Doubleday, 1976.

DYNAMIC SPACE: PROXEMIC RESEARCH AND THE DESIGN OF SUPPORTIVE ENVIRONMENTS [1]

Craig M. Zimring*, Gary W. Evans†, and Ervin H. Zube*

* Institute for Man and Environment
University of Massachusetts
Amherst, Massachusetts

† Program in Social Ecology
University of California
Irvine, California

Frank Lloyd Wright once reportedly described design as "art with a purpose." This description highlights a current view of design. Although the basis of design is aesthetics, design, unlike art, is also characterized by a responsibility to its users. Concern about this responsibility, and with the general topic of human-environment interactions, has prompted a wide variety of design professionals and behavioral scientists to participate in the multidisciplinary research area of environmental behavior. Within the span of a decade, this area has begun to acquire academic respectability as it formed professional organizations, founded journals, and produced many books and articles.

Recently, however, a variety of sources have suggested that the early optimistic expectations for the field have not been realized, especially with regard to the use of behavioral data by design practitioners. Altman (1) provided a thoughtful taxonomy of this

1 We would like to thank Joanne Green, R. Christopher Knight and Harold Raush for their valuable criticisms on earlier drafts of this manuscript.

"applicability gap."[2] In part, he suggested that the problem lies in the analytic approach of researchers which often conflicts with the more holistic synthetic approach of design practitioners. Also practioners claim that their work requires greater specificity and concreteness than is present in most research.

In addition to these problems, however, there also appear to be difficulties in the way in which the environmental behavior field is structured. Due to its interdisciplinary nature, research reports are often scattered, and may appear in journals which are otherwise quite esoteric to the designer. Also, because of a lack of unifying structure it is often hard to see the trend of individual studies, or to predict beyond existing work. Finally, the implications of these studies for environmental design are often not clear.

This paper attempts to address these problems with respect to studies in the human use of space, the area which Hall (26) has called "proxemics." Specifically, two broad areas are discussed: 1) Current trends in crowding, personal space and small group ecology are briefly presented as a basis for later discussion, and several directly relevant studies are discussed in greater depth. 2) A systems approach is suggested as a useful organizing scheme for the design and behavior. The concepts of control, cognitive mapping and arousal are examined as illustrations of this approach. Finally, several resulting design suggestions are proposed.

A BRIEF REVIEW OF TRENDS IN CROWDING, PERSONAL SPACE AND SMALL GROUP ECOLOGY

Crowding

A major debate within the crowding literature revolves around the definition of crowding (Schopler, 50). Freedman (22) and others have defined crowding in terms of density (e.g. square feet per person, people per acre) and have found no consistent ill effects of crowding. Epstein and Aiello (15), Evans and Eichelman (21), Stokols (53) and others have defined crowding as a stressful experience that is associated with the presence of too many other people, and have looked at the various antecedents and the various aspects of this stressful experience. In this view, crowding is a psychological state which must be distinguished from physical density. Stokols (53) has suggested that crowding research can be viewed as the three following fairly distinct approaches: overload,[3] be-

2 EDRA declared "Beyond the applicability gap" as the theme of their seventh annual meeting in Vancouver, B.C., May, 1976.

3 See Evans and Eichelman (1976) for a fuller discussion and critique of the overload model in proxemic research.

havioral constraint and ecological models.

Overload theory suggests that the presence of others may result in excessive social stimulation or information. In this model, people are seen as limited in their abilities to deal with this information and stimulation. When they are pushed beyond their limits, people try to cope and may try to withdraw and otherwise limit input. Milgram (42), for example, followed the lead of early urban sociologists, and suggested that the lack of caring which has been attributed to urban life may result from attempts to cope with an overload of social information (Stokols, 53).

Behavioral constraint studies focus on factors which influence peoples' feelings of control and personal effectiveness. These theorists would predict that a space will be perceived as crowded when the presence of others thwarts important needs (Proshansky, Ittleson & Rivlin, 48). Whether important needs will be perceived as thwarted depends not only on physical density but on such factors as the nature of the group and the nature of the needs. For example, a home may not appear crowded until everyone attempts to shower at the same time in the morning. This crowding may be avoided if the situation is better organized and shower times are staggered, or if the family members are not going to important appointments and hence can avoid showering.

The ecological model is a complex approach which focuses on settings as a basic unit of analysis. The application of this model to crowding results from its emphasis on the relationship of the size of a setting to the experience of the participants. It is hypothesized that every setting has key positions that need filling. For example, the positions in a high school include class officers, athletic team members, and so on. As the size of a setting increases, however, the number of key positions does not increase as quickly as the number of participants. In the former case (when there is a relatively small ratio between participants and key positions), the setting is said to be undermanned. In this situation, people tend to fill many positions, they tend to feel useful to the group, and the group tends to accept strangers fairly readily. Where there is a large ratio of participants to available positions, the setting has been seen as "over-manned," and the opposite experience results; people tend to fill fewer positions, they tend to feel of less use to the group, and the group tends to reject strangers (Wicker, 56). The experience is of being crowded.

A number of empirical studies have indicated that the experience of crowding is related to more than simple physical density. Desor (13) asked people to add stick figures to a scale model room until it "felt crowded." She found that more figures were added in an informal situation (a cocktail party) rather than in a formal one (an airline waiting room). Desor also found that more figures were added when partitions were introduced to the model, and that this effect occurred irrespective of whether the partitions were full-wall, half-wall or transparent. She suggested that partitions had this effect because they reduced social stimulation. Desor's results were replicated by Baum and Davis (3). These researchers found that lighter room color also increased the number of figures placed in their model.

On the other hand, partitions were actually seen to slightly increase the experience of crowding in a real-world replication of Desor's study by Stokols, Smith and Proster (54). Stokols and his colleagues introduced both full-wall partitions and ropes in the waiting area of California driver's license bureau. Both questionnaires and video tape analysis suggested a slight trend towards feelings of greater crowding with the introduction of partitions. The researchers hypothesized that in this setting partitions were seen as herding devices and therefore seemed a nuisance because activities were primarily goal directed. In a more social setting, however, individuals may be more sensitive to social stimulation as a mediator of crowding, and thus partitions may reduce crowding by reducing social stimulation. The disparity between the Desor and the Stokols et al studies also points out that the validity of using scale models to predict real-world behavior is yet to be verified.

PERSONAL SPACE

Personal space (PS) has been defined by Sommer (51) as "the area surrounding a person's body into which intruders may not come." As in the study of crowding, however, the stimulus overload approach has had considerable popularity in studying PS. Hall (26) and many other writers have suggested that as people get closer they gain more information about the people they are approaching. At a very close distance, visual information is supplemented by odor, body heat, and at the closest distance, touch. The large amounts of information available to close distances results in stressful overload. To minimize this stress, people attempt to maintain personal space.

In contrast to the emphasis on information overload as an aspect of PS, Evans and Howard (20) approach personal space from a functional-evolutionary perspective, and suggest that the importance of personal space is to reduce interpersonal aggression by maintaining adequate distance. Altman (1) similarly sees personal space as one of several mechanisms to regulate the level of social interaction for the individual.

The majority of personal space research has attempted to see the effects of personality or experience (called "subject factors" in the behavioral sciences) on the size of the personal space zone. Spatial variables are largely ignored by this research, and hence may limit its usefulness for designers. However, designers are often called upon to design for specific user groups, and in these cases a general understanding of the impact of subject factors is very important.

Evans (21) identified the following six types of variables which influence personal space: 1) Culture. PS is a highly culture-dependent phenomenon, with most data focusing on middle class college-age Americans. Some cross-cultural data indicates that North Americans and Northern Europeans have larger PS zones than the contact cultures of Southern Europe or Latin America. 2) Personality. Smaller PS zones have been associated with a variety of personality traits such as extroversion, less anxiety, positive self concept, and greater change and variety seeking. 3) Abnormality. Individuals with cognitive and personality disorders are more variable in their PS zones than are individuals without such disorders. 4) Sex. Acquainted male-female pairs require less PS than acquainted female-female pairs, who in turn require less PS than acquainted male-male pairs. 5) Age. Mixed findings seem to indicate that children have smaller PS zones than adults, and that onset of adult PS norms occurs at about age 12. Some data indicate that the elderly shift their conversation distance closer than do younger adults (DeLong, 12). 6) Relationship and social situation. People who are well-acquainted tend to interact more closely, and social situations tend to generate smaller PS zones than do formal situations.

SMALL GROUP ECOLOGY

Small group ecology has been defined by Sommer (51) to include studies which deal with the behavior of small social groups as they interact with designed features of the environment. This area is less developed than the study of either crowding or PS, yet is the area which most directly explores the behavioral impacts of design. There are several studies which attempt to relate design features to group social behavior.

Sommer (52) clustered furniture in a geriatric ward in an attempt to increase conversation (sociopetal arrangement), and found a dramatic increase as compared to the previous conditions, when chairs lined the walls (sociofugal arrangement). Holahan (27) replicated Sommer's work in a controlled experiment and found similar results. Mehrabian and Diamond (40) also replicated Sommer's work by using college students and found most conversation occurred when chairs were angled at 90° rather than at 30° or 180°.

Other studies have focused on outdoor behavior. Preiser (47) observed behavior in a public plaza at a university. Undefined open spaces were not used for either social or solitary behaviors, yet each artifact in the plaza had some behaviors that predominated near it. Preiser found standing and talking occurring near columns, leaning near railings, eating near planters, etc. Solitary behaviors were even more tied to these artifacts than were interactive behaviors. These artifacts seemed to provide solitary people a glimpse of the mainstream of activity while providing them some protection. Frequency of behavior decreased further from artifacts, with a rough negative linear correlation between frequency of behavior and distance.

The choice of seating can be a type of territorial signal that can encourage or discourage other users. Sommer (51) observed cafeteria and library users and found that their seating pattern strongly communicated their relationship, and communicated their desire to share a table, as well. Acquainted users would often sit side-by side. A solitary user could discourage later people from sitting at his table by sitting in a position that would force the newcomer to sit close to him, such as the middle seats of a six person table.

SUMMARY

Several approaches to crowding, personal space and small group ecology have been discussed. Crowding has been defined for the purposes of this paper as a stressful experience associated with the presence of too many people. Many factors influence the experience of crowding, such as space, cultural background, group structure and personal needs. Personal space, an apparently related phenomenon, is also seemingly related to many personality and experiential factors, and can possibly be affected by design features such as partitions. Small group ecology is the least developed of the three areas, although it is perhaps the most directly applicable to environmental design. There has been little theory proposed for this area, although creating *opportunity* for conversation (by clustering chairs) and *signaling* appropriate behavior (by clustering chairs and by choice of seating) appear to be key aspects of this research.

We have discussed several theoretical and empirical approaches to proxemic behavior. In the following section we examine a "systems framework" for considering the relationship between proxemics and design.

A CONCEPTUAL FRAMEWORK

If it is true, as was suggested earlier, that a major difference between scientists and design practioners is the analytic approach of the former group as compared to the holistic approach of the latter, then the state of proxemic research may be more comprehensive to designers than to researchers. Designers are traditionally called upon to simultaneously orchestrate the many aspects of design: space, light, color, material, structure. Researchers, however, often look at a very few antecedent conditions and try to link them in a causal way to a few outcome behaviors. In the case of proxemic behavior, the designer's multi-factored approach may be the more useful one. Proxemic behavior is simultaneously dependent on attributes of the individual, and the social situation, as well as on the spatial and designed qualities of the setting.

In a somewhat more formal way, this multi-factored approach may be described as a "systems perspective" (Knight, Zimring, and Kent, 56). Although it is beyond the scope of this paper to fully explore the systems approach, it is necessary to understand some of the general characteristics of this perspective. A system has traditionally been described as a group of elements which form a larger entity. Systems analysis focuses on the elements and their interrelationships and serves as a convenient way to study both the individual elements and the workings of the larger system.

The elements of a system are generally interrelated in complex ways, with each component singly and jointly affecting every other component.

As was suggested earlier, the typical design problem serves as a good example of a highly interrelated system. In designing an office, many elements affect satisfaction. Small spaces provide positive feelings of enclosure, but they may also provide uncomfortable confinement and may communicate low status. Ongoing activity may require quiet or may benefit from ambient noise, and the activity may generate noise, as well. The management structure may suggest spatial arrangements that permit open access to employees or limit that access, or may require variable access at different times. Employee moral may influence the extent to which spaces are used, but may also be affected by the range and quality of available spaces. Bright, primary colors may be pleasurable, but not in offices which are intended to soothe agitated employees. In real-life settings, which are the concern of designers, no single factor determines behavior. Behavior may be viewed as a system which is influenced by multiple factors relating to both the situation and the person.

Moreover, several design and situational factors may jointly affect a "mediating" internal state, which in turn may influence behavior. Knight, Zimring and Kent (56) have proposed the following example where several antecedent conditions (e.g. prolonged social interaction, absence of secluded space) affecting an internal mediating state (need for privacy) which in turn affects several outcome behaviors (e.g. irritability, preference for certain space):

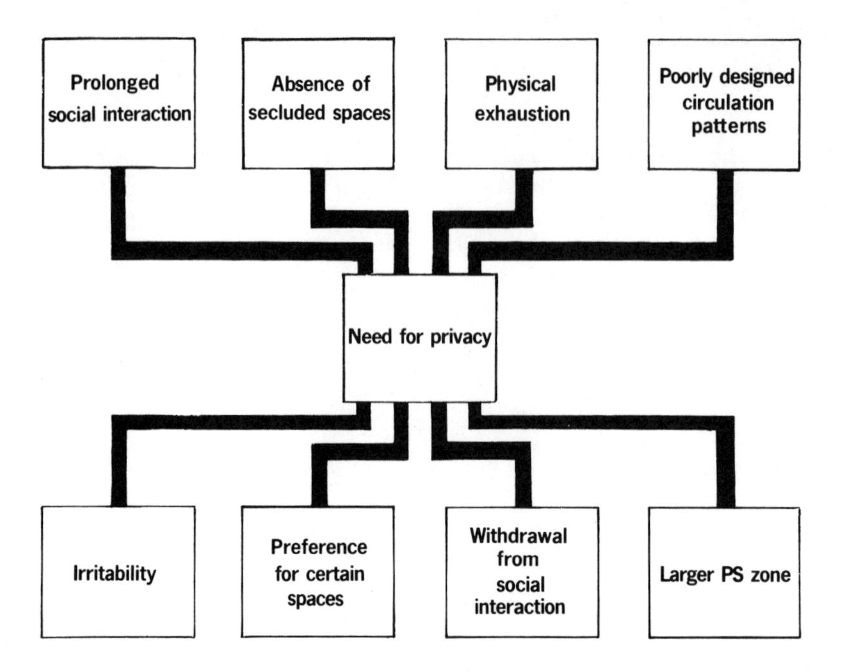

Changes in any or all of the conditions on the left of the diagram may increase or decrease need for privacy, which may then alter the behaviors on the right.

The systems framework for understanding the relationship between the design environment and proxemic behavior is complex. The designed environment is an important element of this system, yet it both affects and is affected by personality-experiential and social-situational factors. Also, this system is better understood if mediating variables are considered; furthermore, understanding how a system operates may require attention to mediating variables.

A general understanding of some of these mediators may be useful to designers in maximizing the positive impact of their designs. Three mediators are discussed below: control, cognitive mapping and arousal.

CONTROL

The concept of control may help to understand the relationship between proxemic behavior and design factors. Design may influence both actual and perceived control, which may, in turn affect a variety of proxemic behaviors. Actual control may be encouraged by a range of design features such as the provision of private spaces with locking doors, or of moveable partitions, such as in Japanese design.

Perceived control may also be fostered by social norms which encourage people to take initiative in dealing with spaces. This link between control, which can be influenced by design, and proxemic behavior has been suggested by some research findings. For example, in crowding research, experimenters following the behavioral constraint model have suggested that an individual's perception of his/her ability to fulfil his/her own needs is a strong predictor of crowding stress (Stokols, 86). Also, in PS research, strangers or individuals in formal situations have larger PS zones than acquainted people, or people in informal situations. Altman (1) has suggested that the PS zone is typically reduced in the latter situations because the individual's perceived control is greater. This would be consistent with Evans' and Howard's (2) notion that a function of PS is to reduce interpersonal aggression. Presumably in the conditions where perceived control is greatest the perceived threat to the individual is minimized.

In small group ecology research, furniture arrangement has been seen to influence conversation with groups which have little control over their immediate environment (e.g. students waiting for interviews, institutionalized geriatric patients). The effects might be less pronounced if these groups felt they had more control. For example, perhaps students in a more informal setting would be less influenced by furniture arrangement, because they feel that they have more control over their environment, which includes the option to move the furniture or, perhaps, to sit on the floor. the behavior of geriatric patients might also be less influenced by furniture if they were encouraged to exert control over their environment. In general, control could be seen as providing freedom of action in a setting. Therefore, the physical setting may be much more deterministic for low power groups than for more potent groups, who may either physically or socially alter the setting.

Knight, Zimring and Kent (56) have suggested that a critical role for the designed environment is to provide opportunity for control. This opportunity can be either actual or symbolic. Providing institutionalized mentally retarded people with lockable doors and private rooms provides them with the opportunity for controlling access to their rooms. Whether this control is exercised, however, depends on whether institution staff respect the residents' privacy. Residents may, in fact, need to be taught how to control their privacy through the appropriate use of their doors.

The use of design to foster symbolic or perceived control is well exemplified in the concept of hierarchy of spaces. This concept is a familiar one in everyday life, and refers to a connected set of spaces which gradually move from individual to group control. For example, we have much control over our bedroom space, less control over the living room, and less control over public street space.

Each space has its own norms and uses. One would expect to see strangers in the street, but not in one's bedroom. A key feature of this hierarch is the gradual transition in control from private to semi-private to public. It would be jarring, indeed, to have a bedroom door which opened directly on the street.

The hierarchy of spaces concept seems useful at several levels of analyses. In his pioneering work, Hall (26) has demonstrated a strong negative correlation between intimacy and distance in many cultures, with specific interpersonal behaviors being appropriate at given distances.

On a larger scale, the usefullness of the concept has been demonstrated in building design. DeLong (12) found in his study of older people that social interaction increased when patients who had formerly slept in a dorm arrangement were given their own rooms (private spaces). It is interesting to note that interaction increased in all areas, both public and private. Patients apparently no longer had a need to stake out a space of their own in the public day room space and conversed in corridors and even invited others into their private rooms. Evidently their perceived control over their rooms was a critical factor. Similar findings were obtained by Proshansky, Ittleson and Rivlin (48) with psychiatric patients and by Esser (16) in his study of disturbed adolescents. Newman's studies of urban crime and proprietorship suggest the importance of a hierarchy of space in both the interior spaces of buildings and in building siting (Newman, 42). He hypothesized that buildings which had a gradation of spaces would foster the positive caretaking behaviors that seems to be lacking in much of our urban life. His correlation data and preliminary quasi-experimental findings indicate that if you provide a hierarchy of spaces, ranging from a private apartment to an area (or stairway) controlled by a few families, to an area controlled by a larger group, and so on, the space defined as belonging to the several families will be protected by them. These areas should be defined by a number of indicators of ownership: physical surveillance, physical access, and symbolic markers such as hedges or ground surface changes. A number of authors, however, have questioned Newman's findings. Greenbie (25) suggested that "distemic" spaces (large scale intercultural spaces) require a very different level of policing than smaller scale "proxemic" spaces because the distemic spaces cannot be subject to small group norms.

An important aspect of the hierarchy of spaces concept involves the fit between control and need for control. It is more critical that we feel control over our bedrooms than over the street in front of our house, and lack of control in the bedroom would probably have more dire personal and psychological consequences. This idea has been formalized by Stokols (53) in his analysis of crowding.

He suggests that environments can be classified as primary and secondary, with important private, personal activities occurring in primary environments such as our homes or offices. Secondary environments include the public environments we encounter every day. Stokols also suggested that needs (and thwarting of needs) can be classifed as personal or neutral. Personal thwartings include situations which are personally important or likely to persist, whereas neutral thwartings are more transitory and less serious. Neutral thwartings in secondary environments are hypothesized to be the least stressful of the four combinations, personal thwartings in primary environments most stressful. For example, we can accept a crowded subway because we know that it's in a secondary environment and that thwartings are likely to be transitory and impersonal. However, if our home is small and poorly organized it may provide a very stressful experience, because important needs must be satisfied there over a long time period.

DESIGNING TO INCREASE CONTROL

Research has suggested that providing control over the environment may reduce crowding stress (hence permitting a larger number of people to comfortably use a space), may reduce necessary PS zones, may increase proprietorship and may help to increase social interaction. At least three techniques might be employed by designers to increase perceived and actual control. These techniques are the following:

1. Designers might provide *opportunity* for control by providing a flexible, manipulable environment which might include private spaces and lockable doors even among institutionalized groups. These options are especially important in primary environments for all user groups. Also, we should provide street furniture in outdoor spaces.
2. Designers might provide symbolic markers to indicate control, by demarkating shifts in control by changes in pavement, paint, floor covering, color or texture. Also, the ability to survey an area visually may increase perceived control over that area (Newman, 43).
3. Designers may help to structure activities in ways to eliminate constraint on the individual, such as by improving communication and circulation patterns. Reduction of constraint through better organization may reduce stress and actually eliminate the need for spatial expansion (Stokols, 53).

COGNITIVE MAPPING

A second mediator which may be useful in understanding the connections between the designed environment and proxemic behavior is cognitive mapping. Cognitive mapping is concerned with how people learn about, organize and remember their spatial world. Work in this field has considered both the processes of developing a cognitive representation of the spatial environment and the characteristics of that representation. The title "cognitive mapping" is a poor one, perhaps, because it suggests the existence of a physical roadmap in peoples' heads. In reality, spatial knowledge is probably coded and stored in a very complex and idiosyncratic way.

Mapping research, like other proxemic research, has focused primarily on subject factors rather than environmental factors. For example, a typical experiment might involve looking at the accuracy of peoples' representation of a town as a function of their experience with that town. Few researchers have tried to assess the impact of changed design features such as added town landmarks on the accuracy of peoples' representation of the town. Lynch (38) would have predicted an increase in accuracy with the addition of landmarks.

Kaplan (33, 34, 35) argues that our very existence calls for quick decisions made on the basis of partial information*, on going "beyond the information given" (Bruner, 8). We must rely on effective information gathering strategies if cognitive maps of our environment are to be developed and useful. Kaplan argues that these strategies for acquiring maps are both "built-in" ("human nature") and learned. That is, both our training and our inherent nature dictate that we *like* to be active, restless organisms who constantly seek stimulation and information. What is more, we prefer the kinds of information that enhances our map (for example, a vista from a tall building). We can deduce human informational needs from examining preferences for different environments. A highly preferred and supportive environment is one that permits us to use the map-building processes of recognition, evaluation, action and prediction. There are examples of such environments in everyone's experience: the mystery of old cities, the tree-lined curving lane that partially obscures what is to come, and the charm of Japanese gardens.

In a series of studies aimed at exploring the process of map-building, Kaplan examined the preference ratings for slides of a number of natural and urban scenes. Complexity theory would suggest that the level of complexity of a scene would predict its preference rating, with moderately complex scenes receiving the highest ratings (c.f. Wohlwill, 57). Kaplan, however, found that five other

*Those who would like to pursue this further should look at the classic text by W. James (31). A more recent treatment is by Bruner (8).

design influenced dimensions of the scenes were also important in predicting preference: 1) mystery; 2) identifiability; 3) coherence; 4) spaciousness; and 5) texture. The slides having a path that wound out of sight, or a grove of trees partially obscuring the view were rated highly. These were scenes that had a *promise of further information*, a sense that somehow if you could move further into the picture you could learn more about the scene. Kaplan's subjects became irate when presented with abstract graphics; if they could not *identify* the scene there was an intensive negative reaction. There was negative reaction for a similar reason when the scene seemed to *lack organization*, when it didn't "hang-together." In a study of scenes from a parkway in Ann Arbor, it was found that views which had at least a modicum of *spaciousness* were preferred. Also, dimensional analysis in the study revealed that observers placed importance on a certain amount of *texture*, which was felt to help define the masses within the spaces as well as define distance (c.f. Gibson, 24).

Lynch, (38) however, in contrast to Kaplan's emphasis on the development of cognitive maps, has focused on the cognitive map itself. In Lynch's view, in order to move around the environment competently, we must have a usable cognitive map. Lynch (38) has pointed to the importance of way-finding for mobile organisms ranging from the polar flights of the artic tern to the path finding of the miniature limpet. Lynch hypothesizes that, in humans, the strategic link between cues present in the environment and effective way-finding is the cognitive environment image, a schematically organized mental map.* Focusing on the urban images which people had constructed of several cities, Lynch found five important common elements in people's sketch maps of cities: paths, edges, districts, nodes and landmarks.

Appleyard (2) used Lynch's technique to explore which qualities of buildings made them effective landmarks in newly designed Ciudad Guayana, Venezuela. Appleyard asked his subject to list remembered buildings, to sketch a map containing those buildings, and to describe a specified route. He found that the distinctiveness of form of the buildings, the visibility from major thoroughfares, the amount of use and the buildings' role in personal activities, and the individuals' inferences about the buildings' cultural significance all affected the degree to which a building was known. Distinctiveness of form accounted for the most variance in people's images, and could be further broken down into seven elements: contour, size, shape, surface, type, quality, and signage. Appleyard suggests that the relative weightings of such elements for other cities might

---

*See Downs and Stea (14) for a treatment of these areas.

be different. His major contribution is the presentation of a taxonomy of causes underlying memory of landmarks, which Lynch suggests are basic elements of urban images.

DESIGNING TO ENHANCE COGNITIVE MAPPING

Although the linkage between design and cognitive mapping is as yet unverified, it appears that it may be possible to enhance cognitive mapping through environmental design. An environment which encourages effective environmental representations may increase effectiveness in way-finding and use of the environment, and thus may also contribute to environments which are preferred by more people. There appears to be at least two ways to increase the map-supporting qualities of the environment:

1. The environment should be imageable. One way to do this is to design and plan using distinctive design features such as landmarks, paths, nodes, edges and defined areas. Areas of perceived control may also enhance imageability. For example, for many people their neighborhood is a discrete and well defined area. This may be true in part because of the perceived control they have over it.
2. The environment should furnish a moderate amount of challenge, where there is "sufficient structure to make sense, to comprehend, to recognize and sufficient uncertainty to make prediction non-trivial." (Kaplan). This would suggest that there should be an overall coherent structure to designs, yet that we should also design for mystery and exploration.

AROUSAL

The theoretical construction of arousal may also connect proxemic research and design analysis. There is considerable evidence that moderate levels of arousal are optimal for a variety of behaviors (Berlyne, 4); (Broadbent, 7); (Kahneman, 32). The function of performance plotted against arousal level is believed to be of an inverted U-shape though the range of optimal arousal is to shift toward lower arousal levels for more complex tasks (Yerkes-Dodson Law). It follows then that either organismic factors or environmental factors (e.g. visual or auditory organization or intensity) which affects arousal level may, in turn, affect an individual's behavior. An individual may seek to control organismic and environmental variables so as to guarantee a moderate level of arousal that is optimal for performance of a given task.

First, let us consider some organismic variables related to arousal. Generally, introverts are more aroused than extroverts (Kahneman, 32). Animal data indicate that subordinate animals have characteristically high levels of arousal (Welch, 35). The extensive work of the Hutts (Hutt and Hutt, 29) has documented that autistic children have abnormally high levels of arousal. Lack of sleep will depress arousal, whereas giving subject's knowledge of their results on tests will increase arousal (Broadbent, 7). Arousal can also be affected by the time of day as well as by various stimulants and depressants (Broadbent, 7); (Kahneman, 32).

With regard to proxemic variables affecting arousal, Epstein and Aiello, (15) have found that crowding significantly increases skin conductance in humans, indicating increased arousal. Comparable data have been found with invasions of personal space (Evans and Howard, 19); (McBride, King and James, 39). Futhermore, Evans and Howard (19) and Evans (18) have found that invasions of personal space and crowding, respectively, only interfere with human performance when the information processing capacities of the individual are pushed to their limits. This pattern of data is consistent with the Yerkes-Dodson Law.

Eye contact has also been demonstrated to elevate arousal, as measured by psychophysiological techniques (Coss, 9); (Nichols and Champness, 44). Links between eye contact behavior and proxemic behavior are also supportive of the proposed arousal mediator. Data generally indicate that reduction of personal space plus the necessity for eye contact is more stressful than reduction of space without eye contact (Evans and Howard, 20).

It is conceivable that persons' preferences for varying amounts of environmental stimulation reflect attempts to optimize arousal. Mehrabian and Russell have found that individuals who are high arousal seekers rate environments that are highly arousing more positively than environments that are not as arousing. The opposite is found for low arousal seekers. If the optimal arousal principle is accurate, one would expect that persons at low states of arousal would be high arousal seekers in order to bring their arousal levels up to the preferable moderate level. The converse would follow as well.

Considerable data also suggest that individuals prefer moderate levels of complexity and may seek and prefer moderate levels of stimulation in the environment (Berlyne, 5);(Welch, 55);(Wohlwill, 56). We believe that the inverted U-shaped performance function, typically found for complexity, is not coincidentally similar to the arousal function. We suggest that high complexity or high stimulation can act to increase organismic arousal whereas a lack of complexity in the environment may reduce arousal. A similar suggestion has been made by Berlyne (6). Findings from sensory deprivation research are not inconsistent with the latter claim. As Wohlwill (58) has pointed out, however, there is little available data concerning the effects of sensory overload.

Nevertheless, some data are suggestive. Hutt and Hutt (28) for example, found that as the complexity of a room increased, the amount of stereotypical behaviors in autistic children also increased. Exposure to a novel environment also caused increases in stereotypic behaviors in autistic children (Hutt and Hutt, 29). Payne (46) has found that for architects there is a highly significant correlation between their pupillary response and ratings of complexity for architectural stimuli. More complex architectural environments were found to invoke greater pupil dilation, which has been suggested as an indicator of heightened arousal. Payne also found that this correlation did not hold for untrained persons; which is consistent with his earlier work (Payne, 45) in which he had found that architects had consistently larger pupillary response to architectural stimuli than nonarchitects. He suggests that this difference between architects and untrained observers may indicate that architects are more aware of, or sensitive to, shifts in complexity in the designed environment than are untrained laypersons.

Payne's hypothesis about the differences between architects and laypersons may reflect differences in previous experience with various dimensions of environmental stimulation. Wohlwill (58) has documented that optimal levels of preferred environmental stimulation reflect individuals' adaptation levels to stimulation primarily as a function of previous experiences with certain dimensions of stimulation. Additional support for Wohlwill's adaptation level hypothesis concerning optimal environmental stimulation comes from some recent work by Friedman (23) who found that people preferred more complex environments if they had differentiated concepts for categorizing physical objects in those environments.

Thus, one's preference for and reactions to various environmental settings may be due in part to the individual's arousal level. This level can be affected by personological, proxemic, and environmental factors. (Greenbie suspects that distemic relationships will reduce higher arousal levels than will proxemic ones, other things being equal (personal comment). Attempts to optimize arousal at some moderate level may be facilitated or inhibited by the configuration of available space. If we are correct in suggesting that organism-environment fit is, in part, a function of arousal, then two hypotheses follow:

First, if crowding and/or personal space invasions increase arousal, then normal persons under such conditions will prefer less arousing or less stimulating environments. Some evidence is available that lends support to the hypothesis. First, Jacobs and Koepell (30) have found that high sensation seekers come from smaller families, presently live singly, and anticipate more mobility in the future than low sensation seekers. Consistent with these data, Cozby (11) reported that persons with a greater desire for change and variety in their physical and social environment tended to have smaller personal space zones. (Personal space zones are typically enlarged with greater arousal.)

A second, more general hypothesis, is that environments which facilitate the reduction of arousal will be preferred by persons who are at high arousal levels. This is of special importance when dealing with highly aroused groups, such as schizophrenics. The converse is hypothesized as well.

Several studies are in agreement with this hypothesis. Berlyne has found that when individuals' arousal levels are increased, persons' tendencies to seek out novel or complex situations diminish considerably. Some data also indicate that introverts, who, you will recall, have characteristically high arousal states, prefer lower levels of stimulation than extroverts (Berlyne, 6).

Coss (10) has recently discussed some of the direct design implications of cut-off or arousal-reducing displacement behaviors. He suggests, for example, that design which reduces the potential for looming*, forced eye contact, and interactions with strangers could help keep arousal at acceptably moderate (optimal) levels. Thus, lowering ambient lighting would reduce perception of other faces, and thus reduce arousal. Traffic corridors could be designed so as to reduce the probability of looming. Finally, distraction or cut-off could be deliberately induced by a variety of methods so as to reduce arousal. Advertisements in subways, fireplaces, windows, and wall decorations may all provide individuals with an easy way to avoid eye contact. Providing objects for people to manipulate could facilitate stereotypical behaviors and reduce arousal as well.

It would be an error for designers and planners to assume on the basis of the above analyses that arousal may provide the sole means for linking proxemic behavior and design. The arousal construct itself is controversial and cannot adequately explain all the data (Broadbent, 7); (Kahneman, 32). Also, arousal is probably not a unidimensional concept as it has been treated here. It is not at all clear whether arousal from one source is qualitatively equivalent to arousal evoked from a different source. We have suggested that perhaps one way to improve organism-environment fit is to provide an environment which helps the organism to obtain or maintain a moderate, optimal level of arousal. We certainly do not believe that arousal is the whole story. Nevertheless, the conceptual power of the arousal construct to organize several previously isolated areas of human-environment studies warrants consideration as an important mediating variable.

---

*Sudden impending collision with a solid body or object which causes a strong increase in arousal and avoidance reactions in many vertebrates.

DESIGNING FOR AROUSAL

As suggested above, arousal may be affected by physical design in combination with many other factors. In most situations a moderate amount of arousal is desired. This would suggest that it is important for designers to be familiar with the overall arousal level of their user group (e.g. schizophrenics are typically highly aroused) as well as the range of arousal levels encountered (e.g. the living environment may be required to operate in a greater range than office environments). Also, the designer should consider the typical social situation likely to occur in the designed space (e.g. calm or anxious) in order to provide designs which help facilitate to a moderate level of arousal. Some ways of influencing arousal are the following:

1. Berlyne (6) has suggested that reduction in size and complexity, the use of the cool colors, and the use of familiarity achieved by redundant design elements may all lower state arousal levels.
2. If we would like to lessen the arousal of having other people in a designed space, we may be able to do so by providing visual foci such as windows, art work or advertising that permit us to avert our eyes from other people in the room (Coss, 9), or by painting rooms light colors to increase apparent size (Baum and Davis, 3).

CONCLUSIONS

We have proposed that a systems approach is an appropriate framework to conceptualize environmental behavior. This framework suggests that the influence of design on behavior is important, but is not deterministic. Design variables must be seen in context of personal and social factors, a number of which have been discussed above. Moreover, this relationship may be seen more clearly by considering mediating variables such as control, arousal and cognitive mapping.

These mediators must be seen as interactive, however, just as personal and social factors are interactive. For example, a more effective cognitive map may increase perceived control, and may reduce arousal. Saegert (49) has shown that subjects' memories of a department store (cognitive map) were poorer in the presence of many people which may have increased arousal. Also, control has been linked to arousal. Introverts, who typically perceive low control over their environment, are often highly aroused (Mehrabian and Russell, 41). Despite these complexities, however, if a designer considers available background information about the users of his/her designs, and considers his/her design in terms of these mediators, a better design should result.

IMPLICATIONS FOR RESEARCH

The richness, complexity and diversity in approach helps to make the proxemics and design field a fertile one for research. Nearly all of the findings that were discussed suggest important further research. Several directions for research seem particularly vital: 1) we must validate our laboratory techniques and findings against real-world situations; 2) we must develop non-verbal measures to supplement easily contaminatable verbal and self-report measures; 3) we must develop multivatiate techniques to examine the interrelationships between independent, dependent and mediating variables. Finally, in all of our research we must continually balance theory and application. While we are acquiring further knowledge about human behavior in real-world contexts, we must keep in mind the question of "how will these data relate to the process of designing more supportive environments in the future?"

REFERENCES

1. Altman, J. *The environment and social behavior.* Brooks/Cole, Monterey, California, 1975.

2. Appleyard, D. Why buildings are known: A predictive tool for architects and planners. *Environment and Behavior.* 1969, *1:* 131-156.

3. Baum, A., and Davis, G. E. Spatial and social aspects of crowding perception, Ms. 1974.

4. Berlyne, D. E. *Conflict, arousal and curiosity.* New York: McGraw-Hill, 1960.

5. Berlyne, D. E. Complexity and incongruity variables as determinants of exploratory choice and evaluative ratings. *Canadian Journal of Psychology,* 1963, *17:* 274-290.

6. Berlyne, D. E. *Aesthetics and psychobiology.* New York: Appleton-Century-Crofts, 1971.

7. Broadbent, D. E. *Decision and stress.* London: Academic Press, 1971.

8. Bruner, J. S. *Beyond the information given.* New York: Norton, 1973.

9. Coss, R. G. The perceptual aspects of eye-spot patterns and their relevance to gaze behavior. In S. Hutt & C. Hutt (Eds.), *Behavior studies in psychiatry.* New York: Pergamon Press, 1970.

10. Coss, R. The cut-off hypothesis: Its relevance to the design of public places. *Man-Environment Systems*, 1973, 417-440.

11. Cozby, P. Effects of density, activity and personality on environmental preferences. *Journal of Research in Personality*, 1973, *7:* 45-60.

12. De Long, A. J. The micro-spatial structure of the older person: Some implications for planning the social and spatial environment. In L. A. Pastalan and D. H. Carson (Eds.), *Spatial behavior of older people*, Ann Arbor: University of Michigan Press, 1970.

13. Desor, J. A. Toward a psychological theory of crowding. *Journal of Personality and Social Psychology*, 1972, *21:* 79-85.

14. Downs, R. M. and Stea, D. *Image and environment: Cognitive mapping and spatial behavior*. Chicago: Aldine, 1973.

15. Epstein, Y. and Aiello, J. Effects of crowding on electrodermal activity. Paper presented at the American Psychological Association, New Orleans, September 2-6, 1974.

16. Esser, A. H. Cottage fourteen: Dominance and territoriality in a group of institutionalized boys. *Small Group Behavior*, 1973, *4:* 131-146.

17. Evans, G. W. Behavioral and psychological consequences of crowding in humans. Unpublished doctoral dissertation. University of Massachusetts at Amherst, 1975.

18. Evans, G. W. Design implications of spatial research. Paper presented at the Seventh Annual Meeting of Environmental Design Research Association, Vancouver, B. C., May, 1976.

19. Evans, G. W. and Howard, R. B. A methodological investigation of personal space. In W. Mitchell (Ed.), *Environmental design: Research and practice*. Los Angeles: University of California Press, 1972.

20. Evans, G. W. and Howard R. B. Personal space. *Psychological Bulletin*, 1973, *80:* 334-344.

21. Evans, G. W. and Eichelman, W. H. Preliminary models of conceptual linkages among some proxemic variables. *Environment and Behavior*, 1976, *8:* 87-116.

22. Freedman, J. L. *Crowding and Behavior.* New York: Viking Press, 1975.

23. Friedman, S. Relationships among cognitive complexity, interpersonal dimensions and spatial preferences and propensities. In S. Friedman and J. Jukasy (Eds.), *Environments: Notes and selection on objects spaces and behaviors.* Monterey, California: Brooks/Cole, 1974.

24. Gibson, J. J. *Perception of the visual world.* Boston: Houghton Miffin, 1950.

25. Greenbie, B. B. *Design for diversity: planning for natural man in the neo-technic environment: an ethological approach.* New York and Amsterdam: Elsevier, 1976.

26. Hall, E. T. *The hidden dimension.* Garden City, N.Y.: Doubleday and Co., 1966.

27. Holahan, C. Seating patterns and patient behavior in an experimental dayroom. *Journal of Abnormal Psychology,* 1972, *802:* 115-124.

28. Hutt, C. and Hutt, J. S. Effects of environmental complexity on stereotyped behaviors of children. *Animal Behavior.* 1965, *13:* 1-4.

29. Hutt, C. and Hutt, J. S. *Behavior studies in psychiatry.* New York: Pargamon Press, 1970.

30. Jacobs, K. W. and Koepell, J. Biographical correlates of sensation-seeking. *Perceptual and Motor Skills.* 1974, *39:* 333-334.

31. James, W. *Psychology: The briefer course.* New York: Harper, 1892, reprinted, 1962.

32. Kahneman, D. *Attention and effort.* Englewood Cliffs, N. J.: Prentice-Hall, 1973.

33. Kaplan, S. Adaptation, structure and knowledge: A biological perspective. In W. J. Mitchel (Ed.), *Environmental Design: Research and Practice.* Proceedings of the Environmental Design Research Association, Conference Three, Los Angeles, California, 1972.

34. Kaplan, S. Knowing man: Towards a humane environment. In R. M. Downs and D. Stea (Eds.), *Image and environment: cogitive mapping and spatial behavior.* Chicago: Aldine, 1973.

35. Kaplan, S. Cognitive maps, human needs and the designed environment. In W. F. E. Preiser (Ed.), *Environmental design research.* Stroudsberg, Pa.: Dowden, Hutchinson and Ross, 1973b.

36. Kaplan, S. An informal model for the prediction of preference. In E. H. Zube, J. G. Fabos and R. O. Bruch (Eds.), *Landscape Assessment: Values, Perceptions and Resources.* Stroudsberg, Pa.: Dowden, Hutchinson and Ross, 1975.

37. Knight, R. C., Zimring, C. M., and Kent, M. J. (In press). Normalization as a socio-physical system. In M. J. Bednar (Ed.), *Physical and social barriers in design.* Stroudsberg, Pa.: Dowden, Hutchinson and Ross.

38. Lynch, K. *The image of the city.* Cambridge: MIT Press, 1960.

39. McBride, G., King, M. and James, J. Social proximity effects on galvanic skin responses in adult humans. *Journal of Personality,* 1965, *61:* 153-157.

40. Mehrabian, A. and Diamond, S. G. Effects of furniture arrangement, props and personality on social interaction. *Journal of Personality and Social Psychology,* 1971, *20:* 18-30.

41. Mehrabian, A. and Russell, J. *An approach to environmental psychology.* Cambridge, Mass.: MIT Press, 1974.

42. Milgram, S. The experiences of living in cities. *Science.* 1970, *167:* 1461-1468.

43. Newman, O. *Defensible space.* New York: MacMillan, 1972.

44. Nichols, K. A. and Champness, B. Eye gazs and galvanic skin response. *Journal of Experimental Social Psychology.* 1971, *7:* 623-626.

45. Payne, I. Pupillary responses to architectural stimuli. *Man-Environment Systems.* 1969, 5-11.

46. Payne, I. Complexity as a fundamental dimension of the visual environment: A pupillary study. *Man-Environment Systems.* 1970, 5-26.

47. Preiser, W. F. E. Analysis of pedestrian velocity and stationary behavior in a shopping mall. Unpublished doctoral dissertation. The Pennsylvania State University, 1972.

48. Proshansky, W. H., Ittleson, W. and Rivlin, L. G. Freedom of choice in a behavior setting. In W. H. Proshansky, W. Ittleson and L. G. Rivlin (Eds.), *Environmental psychology.* New York: Holt, Reinhart & Winston, 1970.

49. Saegert, S. Crowding: Cognitive overload and behavioral constraint. In W. F. E. Preiser (Ed.), *Environmental Design Research.* Vol. II, Stroudsberg, Pa.: Dowden, Hutchinson & Ross, 1973.

50. Schopler, J. Conceptions of crowding. *Science.* 1976, *192:* 64.

51. Sommer, R. *Personal Space: The behavior basis for design.* Englewood Cliffs, N.J.: Prentice-Hall, 1969.

52. Sommer, R. and Ross, H. Social interaction on a geriatric ward. *International Journal of Social Psychiatry.* 1958, *4:* 128-133.

53. Stokols, D. Crowding in primary and secondary environments. *Environment and Behavior.* 1976, *8:* 78-87.

54. Stokols, D., Smith, T., and Prostor, J. Partitioning and perceived crowding in a public space. *American Behavioral Scientist.* 1975, *18:* 792-814.

55. Welch, B. Psychophysiological response to the mean level of environmental stimulation: A theory of environmental integration. In *Symposium on Medical Aspects of Stress,* Walter Reed Army Institute of Research, 1964.

56. Wicker, A. W. Undermanning theory and research: Implications for the study of psychological and behavioral effects of excess populations. *Representative Research in Social Psychology.* 1973, *4:* 105-206.

57. Wohlwill, J. F. Amount of stimulus exploration and preference as differential functions of stimulus complexity. *Perception and Psychophysic.* 1968, *4:* 5.

58. Wohlwill, J. Human adaptation to levels of environmental stimulation. *'Human Ecology.* 1974, *2:* 127-147.

CONTEXT, STRUCTURES AND RELATIONSHIPS

Alton J. De Long

School of Architecture
University of Tennessee
Knoxville, Tn. 37916

The theme of this book is the *complementary relationship* between privacy and communality. Entities which are complementary, like the two sides of a coin, are merely different manifestations of the same thing: together they constitute an essential unity. To speak of privacy and communality as complements, however, involves more sophistication than simply acknowledging that both are important for the individual or the group. Complements are not opposites, although too often they are treated that way. Complements are mutual contributors to an underlying cause, or to an overall picture, much like the pieces of a jigsaw puzzle each of which uniquely contributes to the unitary integrity of the puzzle.

Privacy and communality will be viewed as complements with respect to the intelligible organization of social relationships. The 'puzzle' involved, or the overall 'picture' we want to comprehend is *the regulation of interpersonal access* with the result that the framework employed does not treat privacy and communality as oppositions; but rather as a continuum with alternating states possessing a variety of degree of manifestation.

The complementary relationship, as we will show, is central to understanding the intricate connections between environment and behavior. Unfortunately, these types of relationships are typically ignored in man-environment research because they demand a careful accounting of the role of *context*. As Edward T. Hall (22:86) has stated, most ". . .research in the social and biological sciences has turned away from context. In fact, attempts are often made to consciously exclude context." The type of logic which will be employed in this paper, therefore, will be somewhat different from that customarily relied upon by social and behavioral researchers, although certainly not any more complicated: indeed, every normal human being has mastered it by the age of six.

Dealing with complementarity raises a most fundamental issue especially pertinent to the field of man-environment relations. When dealing with complements there must be an underlying unity or totality involved. Many times the nature of the totality is not known -- only the complements can be observed and the underlying unity must be inferred. This problem is potentially confusing unless the logic employed is contextual in nature. As normally functioning human beings, we all employ contextual logic thousands of times every day as we regulate our relationships to others. A simple smile, for example, cannot be treated independently of its context because the same smile in one situation can mean something entirely different in another. Similarly, different smiles can possess essentially the same meaning if they occur in different situations. In fact, unless the context is known, it is often extremely difficult to determine laughter from crying from the visual image alone.

Artists, of course, must learn the lessons of contextual logic very quickly. Alizarin crimson does not have the same color value in a field of green as it does in a field of bright blue or intense yellow. The value of a color changes dramatically when placed in the context of other colors, often much to the dismay of the amateur painter. Designers, too, are well aware of the role of context. A twelve foot ceiling can be perceived to be very high or rather low depending upon where it occurs, while a 450 square foot dwelling unit can seem hopelessly cramped or luxuriously spacious if it is contexted appropriately.

The underlying totalities in the above examples deal in the first instance with the fact that a variety of smiles can be semantically and functionally equivalent; in the second instance, with the fact that a variety of crimsons and shades of pink can have the same perceptual value; and in the third instance, that ceiling height and area also vary according to the context.

The value of observed events, then, whether facial expressions, colors or space is not absolute, but rather always a function of the environment in which they occur. In other terms, the totality or perceived unity which emerges when concerned with complements-in-context is a *relationship*.

Man-environment relations is an implicitly contextual statement; and it is accurate to say that the emergence of the field itself was the result of growing dissatisfaction with noncontextual orientations toward the study of man and his behavior. This essay is organized to elucidate why a contextual approach is specifically suited for dealing with the issue of privacy and communality; and, in general, why it is necessary for the study of man-environment relations.

This essay will deal first with empirical illustrations concerning the structuring of microspatial environments, showing why the application of contextual logic is essential if the behavioral implications of microenvironments are to be accurately understood in terms of whether people perceive themselves as being "together" or "apart" in a given space. Attention will then be directed to the issue of structured relationships and the regulation of boundaries in order to derive the complementary nature of privacy and communality. Selected examples in the regulation of interpersonal access will include the manipulation of spatial, social and conceptual boundaries. Since the perceived or functional unity arising from complements-in-context resides in relationships, we will examine the curious and crucial properties of such relationships. These relationships introduce stability and exhibit properties of invariance even though they rely upon continually varying components for their realization -- much like the fractional relationship "one-half" maintains an invariant identity even though its realization involves continually changing components, i.e. 1/2, 2/4, 3/6 etc. We will also examine the seemingly paradoxical property of such relationships; namely, that the greater the intrinsic diversity within a system, the greater the potential manifestation and expression of unity. We will then briefly consider the characteristics of contextual logic and show that the principles involved are not esoteric, intellectual abstractions; but rather seem to be the result of how our central nervous system is organized. To continue to ignore the parameters of this basic logic seems equivalent to maintaining the dualistic, "either/or" encapsulation of the relationship between man and his environment. If privacy and communality are viewed as competing oppositions, an inevitable conclusion is that one will ultimately be more desirable than the other. If they are viewed as complements, both are requisite for the full expression of human potential. Contextual logic is not exclusionary or competitive, but systemic and synergic.

MICROSPATIAL STRUCTURES AND INTERPERSONAL ACCESS

Since the term structure has so many different connotations, it will be well to indicate the current use of this term. When we refer to structure we are referring to constraint. A set of events can be said to be "structured" if the occurrence of any particular event has implications for the appearance of another. If the occurrence of an event has no bearing on the occurrence of any other event, and "anything" is equally likely to happen, then, the set of events are said to be random in their occurrence and lack structure. The most straightforward definition of structure is that provided by Gleason (17:382):

> "Structure is merely a set of limitations on freedom of occurrence and hence inevitably produces redundancy."

As Von Foerster (31) has illustrated, the number of constraints which must be introduced to produce regularity and predictability is surprisingly small. The structural properties of behavioral systems have several crucial aspects (De Long, 9) which cannot be elaborated here. It will be sufficient to say that relationships (units) derived from complements-in-context are one aspect of structure while their modes of combination constitute another.

The presence of structure is what makes it possible to learn something as complex as language in a relatively short period of time without being consciously aware that you are doing so; and it is predictable structural relationships, or pattern, which permits one to manipulate the structure of a system and in the process create totally new expressions which are mutually intelligible (for example, "uncola").

While structure results in predictability and expectation, it is not possible to predict the structure of one system either *a priori* or based upon experience with another system. This is because structures in behavioral systems vary from culture to culture, subculturally and often even within smaller reference groups: the structural properties of behavioral systems are arbitrary, or irrational (Hall, 19, 21, 22).

This point was made abundantly clear four years ago when two design students were asked to design a space conducive to relaxing, comfortable and casual social interaction. The task initially seems simple enough, but other constraints were imposed; namely, the designers had to *predict* how the space would be used. To make the predictions, they used themselves as controls. To their genuine surprise, their predictions were hopelessly inadequate.

The designed environment is shown in Figure 1a, consisting of two soft armchairs, a full sofa, two floor lamps and a round coffee table.[1]

---

1. All work on microspatial structures was conducted in scale-model environments (1" = 1'-0") unless otherwise specified. For papers dealing with the reliability and validity of research conducted at this scale, see De Long (12, 13, 14).

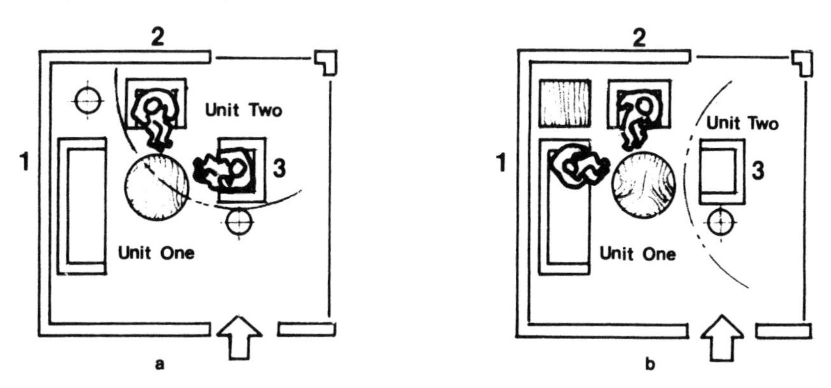

Figure 1: Micro-spatial segmentation and seating selection for friends.

When the designers of the environment tried to predict where users of the environment would sit to carry on a comfortable, casual conversation with a friend, they occupied the seating positions which without exception *were selected by informants as the most comfortable positions for themselves and a stranger with whom they did not care to interact!* The situation was puzzling to say the least since previous research with similar arrangements had all indicated that "inside" positions (those selected by the designers) were the ones selected for the interpersonal relationship specified, i.e. casual conversation between friends. Additional subjects were employed, but again with exactly the same results. What we had perceived as being "together"; namely, the sofa and the chair, was apparently being perceived by subjects as being "apart." That is, they always occupied positions 1 and 2 for relationships in which they wanted to limit access to themselves (a stranger with whom they did not want to interact), the opposite of what we had expected. Separation implies a boundary, and so we began to analyze the situation, and try modifications. When we arrived at the arrangements shown in Figure 1b we had a fairly good idea of what was happening, and called the same informants back. Their response was, "Oh, this is different from before," and they occupied seating positions 1 and 2 for a conversation with a friend, as originally anticipated.

What is involved in this example is the differential segmentation of microspace based upon underlying structural units which differ in the two settings. The situation is very similar to the difference between "Ho Ping" and "Hoping." Ho Ping, my chinese friend and Hoping (something good will happen) elicit as contexts quite different behavior from me. While containing similar underlying elements, they are *segmented* differently: "Ho + Ping" versus "Hop(e)+ ing." And the difference in how ostensibly similar ingredients are structured *completely* shifts the behavioral interpretation of the individuals' location and orientation within the environment.

It should be kept in mind that the selection of positions 1 and 2 in Figure 1a *restricted* interpersonal access while the selection of those positions in Figure 1b *permitted* interpersonal access.

This example is important because it illustrates two fundamental principles: namely, that the same spatial behavior does not possess a constant value (or meaning), and that different spatial behavior can possess the same value. For purposes of illustration, we will present these two principles graphically and formulaically. The formulaic representation will be referred to later in order to demonstrate that these principles apply to a variety of phenomena across a variety of levels of complexity. To translate the graphic into the formulaic we will say that the location and orientation of individuals in seating positions 1 and 2 = a_1, and that the location and orientation of individuals in positions 2 and 3 = a_2.

Thus, we have,

or, if the same event occurs in two different contexts, the value of the event will not be equivalent, or constant.

Formulaically,

$$a_1(C_1) \neq a_1(C_2)$$

and,

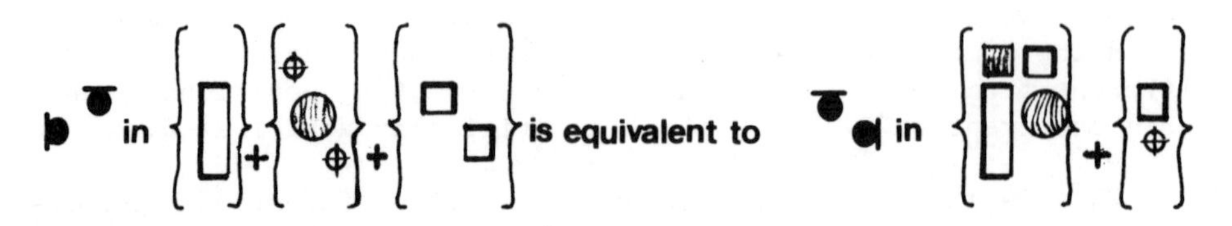

or if two different events which can be considered similar occur in different contexts, the different events may be functionally equivalent.

Formulaically,

$$a_1(C_1) = a_2(C_2)$$

While such segmentation between the underlying structural units of microspace is indeed subtle, the behavioral effects are pervasive. However much the two spaces may appear to be similar, they are perceived as being radically different structurally; and in order to maintain the same interpersonal relationships, the spatial behavior of informants undergoes systematic alterations. In this example we see the same primary artifacts as being interpreted as being "together" in one instance and "apart" in another, *even though the distance between them remains unaltered.* In short, the relationship "casual conversation with someone you know" occurs *within* the same microspatial unit, whereas the relationship, "no conversation with a person you do not know" occurs *across* unit boundaries.

Pursuing the relationship, "casual, comfortable conversation with someone you know," a bit further, we can identify how the size of the setting and the orientational relationship between artifacts function as contextual factors for spacing itself.

Robert Sommer (28:66) reports a simple study which attempts to identify the "limits of a comfortable conversation." The study involves the use of two sofas in a head-on orientation, or facing one another directly. Sommer's findings indicated that people would sit opposite one another until the sofas were 3'-6" apart, at which distance they would begin to sit side-by-side on the same sofa. Scale-model study of the same situation yielded precisely the same results (De Long, 12). Additionally, the scale-model study incorporated a condition to examine a suggestion made by Sommer that a shift in setting would alter the results. Specifically, Sommer observed that, ". . .as room size becomes larger, people sit closer (Sommer, 28:67)." Scale-model study verified Sommer's observation. In a smaller setting, people sat opposite one another on separate sofas until a distance of 4'-0" was reached.[2] The overall size of the setting as well as the microspatial structuring of an environment, then, affects the functional value of distance per se. We turn now to an example which will illustrate that orientation also nullifies attempts to treat any specific distance in an absolute manner.

Orientation between interactants is a prime consideration in proxemic analysis (Hall, 20). Orientation does not, as is sometimes thought, refer to eye behavior; but rather to the relation-

2. A distance of 4'-0" between sofas yields a "nose-to-nose" distance of 6'-6" when appropriate dimensional adjustments are made. This distance, of course, constitutes the boundary between "social-close" and "social-far" transactions (Hall,21: 114-115).

ship established between interacting individuals when lines, or planes, are constructed through their shoulders. It is thus an angular relationship which indicates the degree to which interactants' bodies are facing one another (regardless of whether or not they are looking at one another). Body orientation as outlined by Hall is illustrated in Figure 2.

Scale-model research which has replicated the spatial zones outlined by Hall for seating within furniture arrangements has additionally indicated that the proxemic zones display systematic variation depending upon the orientational context (De Long, 14). Table One shows the proxemic zone midpoints which were empirically identified across three orientations. Table Two illustrates the patterning involved. In Table 2, letters are assigned to distances such that the same letter always refers to equivalent distances. "B", for example, represents 4'-8", 4'-9" and 4'-10", all of which are considered equivalent to one another. Table 2 illustrates, in the first instance, that the head-on orientation always employs the largest distance while the side-by-side orientation always employs the smallest distance for any given proxemic zone. For zones 1-3, the right-angle orientation employs the same distance as the head-on orientation while in zones 4-7 it employs the same distance as the side-by-side orientation. It should be noticed that the largest distance in one zone becomes the smallest distance in the next zone, with the exception of zones 4 and 8.

Second, it should also be noted that Table 2 shows that *different distances* can promote the *same degree of interpersonal access*, and that the *same distances* can promote *different degrees of interpersonal access*. In the former situation within zone 1 we see that 3'8" and 4'9" (a & b) both promote "Personal-Far" transactions, "a" occurring with side-by-side orientations and "b" occurring for head-on and right-angle orientations. With regard to the same distance promoting different degrees of interpersonal access, we can see that "b" occurs in zone 1 with head-on and right-angle orientations and in zone 2 with the side-by-side orientation. Distance "b", therefore, signals "Personal-Far" transactions with two orientations, but "Social-Close" transactions when it occurrs with the third orientation. While this may appear paradoxical, it is not because the various distances which signal the types of interpersonal trans-

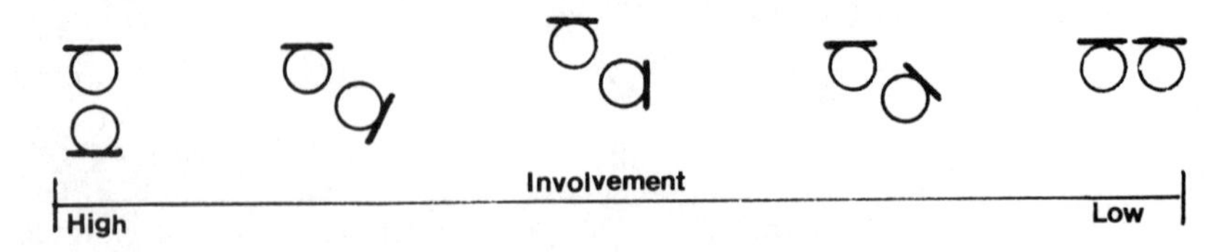

Figure 2: Body orientation and interpersonal involvement

TABLE 1:

DISTANCE-VARIANTS* (ALLO-PROX') ACROSS ORIENTATIONAL CONTEXTS

	Proxemic Zones (Texas-Scale)							
Orientation	①	②	③	④	⑤	⑥	⑦	⑧
[ ]	4^{10}	5^{11}	7^{1}	10^{7}	14^{11}	19^{1}	21^{10}	26^{3}
[⊔	4^{9}	6^{3}	7^{1}	8^{11}	11^{0}	13^{9}	19^{2}	25^{6}
⊔ ⊔	3^{8}	4^{8}	6^{1}	8^{8}	11^{4}	14^{2}	18^{9}	23^{10}

*Distances are "nose to nose" based on zone midpoints

TABLE 2:

PATTERN OF ALLO-PROX DISTRIBUTION

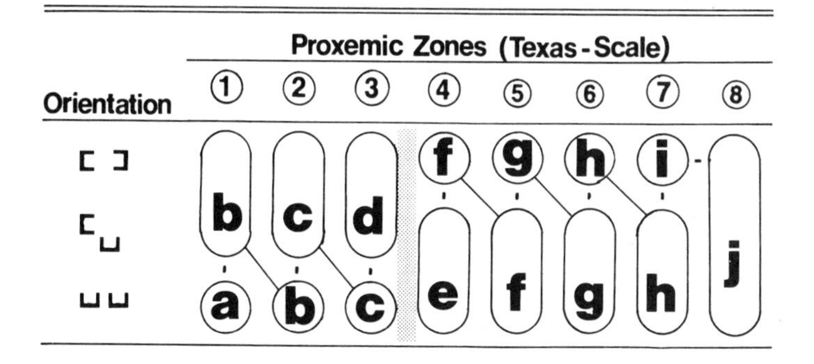

actions are systematically distributed across the different orientations. Thus distance "b" is the *complement* of distance "a" when it occurs in either head-on or right-angle orientations, but functions as a *contrast* with distance "a" when it occurs in the side-by-side orientation (i.e., it constitutes a difference-which-makes-a-difference). The systematic distribution of distances across the orientations for the proxemic zones means that instead of having to deal with 24 distances (8 zones x 3 orientations) to regulate interpersonal access, only 10 are required, a considerable reduction of complexity.

Of the eight zones shown in Tables 1 and 2, zones 1, 2, 3, 4 and 7 are made by all informants while zones 5, 6 and 8 appear to be regional distinctions made by Texans. Taking the zones all informants have in common, we see that within furniture arrangements, there are five zones employing 8 distance-variants. If we treat the proxemic zones as units with respect to the degree of interpersonal access, we can define a proxemic zone as a unit-of-space in the following manner:

> A proxemic zone constitutes a spatial unit consisting of two principal distances (variants) each of which occurs in a unique (i.e., complementary) orientational context.

Letting units be designated by brackets, { }, and the variants be designated by / /, we can precisely define the proxemic zones by specifying the complementary distribution of the variants which constitute each zone. Thus,

"Personal-Far" is a unit,
{D_1}, consisting of /3^8/ in context, ⊔⊔
alternating with /4^9/ in contexts, ⊏⊐ ⊏⊔

"Social-Close" is a unit,
{D_2}, consisting of /4^9/ in context, ⊔⊔
alternating with /6^0/ in contexts, ⊏⊐ ⊏⊔

"Social-Far" is a unit,
{D_3}, consisting of /6^0/ in context ⊔⊔
alternating with /7^1/ in contexts, ⊏⊐ ⊏⊔

"Public-Close" is a unit,
{D_4}, consisting of /10^7/ in context, ⊏⊐
alternating with /8^8/ in contexts, ⊏⊔ ⊔⊔

"Public-Far" is a unit,
{D_5}, consisting of /21^{10}/ in context, ⊏⊐
alternating with /19^0/ in contexts, ⊏⊔ ⊔⊔

It should be apparent that degree of interpersonal access is not a function of distance per se, but rather is a *function of the manner in which a given distance is structured within the micro-spatial system.* A distance of 6'-0" may be functionally equivalent to 4'-9" or 7'-1" ($\{D_2\}$ and $\{D_3\}$) depending upon the orientational context in which it occurs. In other words, the employment of exactly the same distance of 6'-0" can be manipulated to encourage different degrees of interpersonal access ($a_1(C_1) \neq a_1(C_2)$) depending upon the context. Similarly, the use of different distances (4'-9" and 6'-0") can result in the same degree of interpersonal access ($a_1(C_1) = a_2(C_2)$). It is obvious that structures and their manipulation are central to an understanding of interpersonal access, and to the issue of privacy and communality.

STRUCTURES AND BOUNDARY-REGULATION: THE ORCHESTRATION OF INTERPERSONAL ACCESS

A recent conceptual analysis of privacy by Altman (1) will provide a point-of-entry into our consideration of the relationship between privacy and communality. Altman's analysis is interesting from several viewpoints. First, he defines privacy in very general terms, saying it is, ". . . selective control of access to the self or to one's group (Altman, 1)." Additionally, Altman postulates certain characteristics of privacy which include, ". . . linkage between combinations of persons or groups," (1); a dialectic nature which involves the selective opening and closing of boundaries (1); the regulation of boundaries (1); and, ultimately by extrapolation, the control of input and output.[3]

Altman's analysis leads to a rather simple conclusion: privacy has something to do with *the regulation of what happens between people.* Those who have carefully studied processes in the regulation of what happens between people typically refer to such processes as *communication.* Anyone who has seriously examined the communication process empirically soon realizes that what occurs between people is regulated, or structured, in a variety of extremely subtle ways. If privacy can be considered related to the regulation of boundaries, the central issue in such regulation is *structure;* and it is upon the role structure plays in boundary-regulation that we will now focus our attention.

Our illustrations in this section will be formulated within the framework suggested by Esser (16:118-119) which postulates that man structures his relationships to his environment physically,

3. Altman includes several other properties in his analysis, but they are irrelevant to the discussion we will present.

socially and conceptually. Man, then, inhabits three distinctly different types of spaces: physical, social and conceptual. While distinct and apparently the result of different neurologic substrates (MacLean, 25), these spaces are well integrated with one another and are not to be considered independent of one another. Man structures his relationships to others on all three levels simultaneously, often the different levels reinforcing one another, but occasionally levels may contradict one another in which case it is important to know which level is primary in a given situation. Additionally, these three levels are hierarchically organized, the evolutionary emergence being initiated at the physical level, followed by the social and then the conceptual. We will begin with how we structure our relationships to others at the most basic level, physical space.

The importance of structure becomes apparent when the underlying elements in a situation are held constant and the structural relations between the elements altered to produce different effects. The four microspatial environments shown in Figure 3 exhibit this property: the underlying elements are exactly the same, but all four environments are structured differently. Because each of the microspatial environments is structured differently, the *units* involved in the construction of the environments differ, as do the microspatial boundaries. As a consequence, the relative degrees of access to person "A" differ, until as shown in Figure 3d, no one has access to one another.

To say that "no one has access to one another," requires qualification in the sense that *according to the microspatial structuring access is not permissible* Interpersonal access at the level of microspatial structure requires that individuals be situated *within the same spatial unit.* In Figures 3a and 3b individuals "A" and "B" are within the same spatial unit, while in Figure 3d they are in different units. When individuals are situated in different units, interpersonal access to one another must typically be accompanied by a higher-order of structuring which overrides the microspatial organization. For example, a person wanting access to another might initiate postural synchronization (Scheflen, 27) --

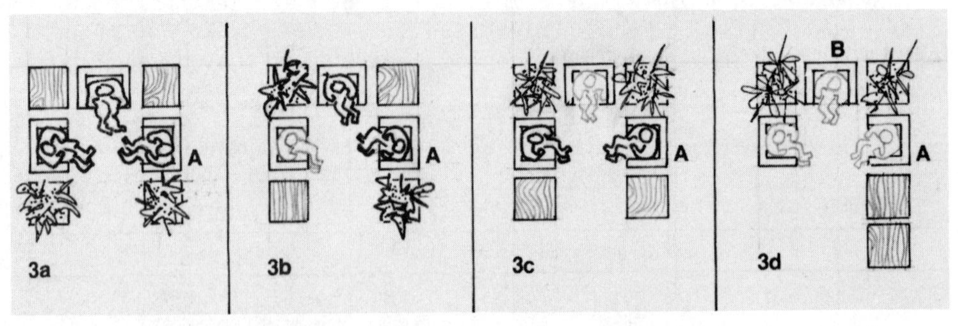

Figure 3: Interpersonal access as a function of micro-spatial structuring

a sharing of postures indicating 'togetherness' -- or initiate a neutral, linguistic-conceptual structuring such as, "Nice day, isn't it?" Of course, if "A" is being approached by higher-order structuring on the part of "B" in Figure 3d, and wants to avoid interaction he can refuse and rest assured that the microspatial structuring will reinforce his decision.

It should be clear in Figure 3 that boundaries are involved and that the regulation of those boundaries is a function of how the space is structured. It should also be clear that how a space is structured has implications for whether or not the same density will be perceived as being tolerable or potentially stressful -- compare 3d where each person is structurally separate with 3a where all occupants are structurally together and have access to one another.

In the foregoing example, the term boundary-regulation is simple enough to grasp -- we are speaking of spatial boundaries. But the notion of boundaries in the sense of "the demarcation of one thing from another" is central to the understanding of any communication process, as we will illustrate with kinesic communication between children.

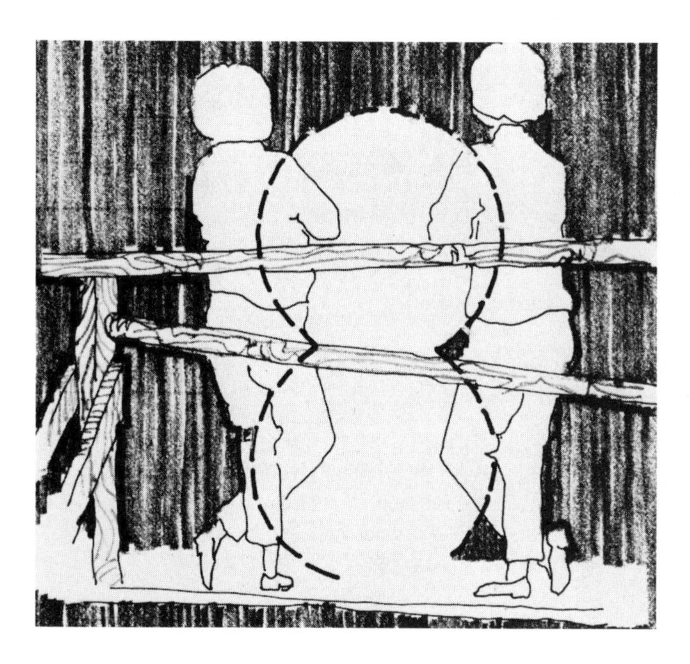

Figure 4: Postural synchrony - a common form of social sharing communicating "togetherness"

Adults often complain that children around four and five years of age continually interrupt conversations. Our empirical observations of children interacting among themselves did not support the popular stereotype. As a matter of fact, it was found that the *regulation* of speaker-listener roles was remarkably patterned and efficient. How the regulation is managed required two years of intensive microkinesic observation, but turned out to be incredibly simple.[4] The problem emerged as a boundary-regulation problem: How does one child of four years old know when another is intending to conclude speaking and is willing to give him access to the speaker's role? The exchange of speaker and listener roles involves identifying a kinesic structure which alerts one to the presence of the boundary between when he can stop listening and begin speaking. The kinesic structure is crucial because the boundary is not always marked linguistically. Our analyses indicate that termination signalling (i.e. signalling to one's partner an intention to yield the speaker's role) is marked by a mandatory leftward movement in the head of the speaker accompanied by downward movements which may or may not (depending upon context) occur in the head. Further, these left and down movements must be tightly clustered and occur in at least one of three positions within the utterance: the penultimate word, the final word or postverbally. The variant forms of the signal differ depending upon the utterance position in which they occur. Most important is how *the display or lack of display* of this structural unit *controls* access to active conversation.

Normally, when a child concludes an utterance and is willing to exchange his speaking role for that of listening, the signals are kinesically emitted and the exchange is accomplished smoothly and for the most part totally unnoticed by outside observers. The two examples we will discuss however, highlight the *degree of control of access* to the conversation that such structures exert. In the first instance, a child says,

"I'll show you what's brown . . ."

After this utterance there is an 8 second silence, or delay, during which neither child speaks. Now say the above utterance, and then count out eight seconds -- it is an incredibly long silence in a conversation. After the eight second delay, the child who was speaking (and still had the 'floor') moved his head *left* and *down* and his left hand *left*. Immediately thereafter the other child began speaking. He could not assume the role of the speaker until the appropriate *structure* which would tell him it was permissible to

4. For a detailed treatment see, De Long (11). For a summary, see De Long (15c).

exchange roles had been emitted. Lacking the appropriate structural display a boundary did not exist across which the speaker and listener roles could be exchanged. By witholding the termination signal, the speaker controlled the access of the listener to the speaking role by, in effect, refusing to create a boundary where exchange could take place.

In a second example, we find the use of the same structure employed in a manner which invites the listener to enter the speaking role in the *midst* of an ongoing utterance. One child says,

"And this, this is an upside-down cake. Uh, that's a puttiy cat. And this,"

Right after the utterance, " . . . And this . . ." the child emits a *left* in the head, a *down* in the head, and a massing of downward movements throughout the left arm and hand. Immediately, the listener enters the conversation with,

"I got a puttiy cat, too."

The child who "interrupted" the conversation *knew* she was in the speaker's role by invitation and immediately upon concluding her utterance signals the other to continue speaking by emitting the left-down structure. The kinesic behavior of the "interruptor" was analyzed frame-by-frame on sound-motion film (24 fps) and is markedly constrained in this instance compared to her behavior when normally occupying the speaking role. What is particularly impressive is that this "interruption" takes place perfectly smoothly in a time-span of less than one second, and the original speaker resumes her conversation as though nothing had happened at all.

In this second example, the manipulation of kinesic structure created a boundary at which the exchange of roles could be conducted (the listener was permitted access to the speaker's role) smoothly and efficiently; and further the context in which the structure was manipulated informed the listener her access to the speaker's role was to be of short duration only (i.e. "limited access"). Throughout the research conducted on termination signallying (intention to terminate the speakers role) in preschool children, the most impressive observation was how predictably and precisely access into and exit from the speaker's role was regulated between interactants through the manipulation of this aspect of kinesic structure; and further how binding the manipulation of structure was on both parties involved in the transaction.

It is essential not to lose sight of the fact that structural units are in every sense of the term *units*. Units are differentiable from one another. That is, they begin and end; and it is where one unit ends and another begins that boundaries exist. *Boundaries are implicit in units; and boundary-regulation is ultimately synonomous with the manipulation of structures.* When a user of a system perceives units he also perceives boundaries and vice versa. The regulation of boundaries, then, is the regulation of structures.

We have outlined how one moves from the role of listener to the complementary role of speaker within a conversation. Getting into a position where one can play either the role of speaker or listener is an equally complex affair; and one based, it appears, upon temporal structures such as rythms. In other words, to get into a conversation or to have access to being a party to a conversation is not the casual affair many would believe. There are suggestions that to have access to a conversation one must "enter" (as listener or speaker) upon the stress beat of the speaker; and that if this is not accomplished it is impossible to have access to the conversation even though standing beside those who may be involved in the conversation. What this suggests is that one who can continually shift his timing system can through temporal structuring alone effectively limit access to a conversation (A. R. Esser, personal communication) and thereby achieve an effective degree of isolation from others directly beside him.[5]

Moving to the conceptual, symbolic level of structures we will take as an example, linguistic style as outlined by Martin Joos (23). Native speakers of American English employ five distinctly different styles of speaking which are characterized by structural, or code-features, involving phonology, morphology, syntactic and semantic variations. These styles are employed systematically, and their usage and sequential manipulation clearly signal to the listener not only his degree of involvement in the conversation, but the degree of access he will be permitted and the form his access must follow.

Joos (23:11) specifies five basic styles: "Intimate," "Casual," "Consultative," "Formal," and "Frozen," each of which informs conversational participants as to the type of information which can be permissably supplied during the conversation as well as the informa-

---

5. Ada Reif Esser's work with rhythmicity leads her to the theoretical conclusion that the manipulation of temporal structures is one of the crucial ways in which individuals regulate their sense of privacy (personal communication). William Condon has analysed a film of a mother and twin daughters during which one twin is consistently denied access to the conversation through systematic shifts in postural synchrony by the mother and the other daughter.

tional access and demands appropriate during the transaction. The features of each style will be briefly summarized.

"Intimate" style involves the understanding between participants that no public information or background will be provided; and that the addressee is expected to know precisely what is meant by "filling in" or "supplying" the appropriate context. The message, then, is the product of both parties to the transaction. Each party to the transaction has total access to the other. As Joos (23:29-30) remarks, the speaker " . . . avoids giving the addressee information from outside the speaker's skin." The style is characterized by brevity. It is the high-context communication Hall (22) refers to, and the restricted code Bernstein (4) outlines.

The "Casual" style is reserved for " . . .friends, acquaintances, insiders: addressed to a stranger, it serves to make him an insider . . ." (Joos, 23:23). Casual style assumes that background information and advance programming is unnecessary since "being programmed" is what it means to be a legitimate insider. The employment of the casual style is a signal that the participants are members of a *social group*. It contrasts with the Intimate because information from **outside the speaker's skin** is conveyed and it contrasts with the Consultative style because the Casual style takes public information for granted.

The "Consultative" style" is characterized by two features: "The speaker supplies background information . . . (and) the addressee participates continuously" (Joos, 23:23). The speaker does not assume he will be understood without supplying the appropriate background, and the listener is obligated to continuously acknowledge (verbal acknowledgement is required about every six seconds) he is receiving the background information and understands it. It is the style which is typical of transactions with strangers -- those we can safely assume are speaking our language, but who may have different background experiences. It should be obvious that the listener's access to the speaker is of a different nature in this style than with those discussed previously. It is also interesting to note that the consultative style can be maintained for a group size of up to about six individuals, after which the style must shift to the "Formal."

"Formal" style, according to Joos, differs from Consultative in that participation, "drops out." The purpose of the Formal style is to inform; and as a consequence the background information is woven throughout the discourse in complex patterns which are enunciated carefully and articulated according to the dictums of fine logical relationships. The manner in which participation "drops out" on the part of both speaker and listener is extremely interesting. In the Consultative style, the addressee must participate insofar as acknowledging the reception of background information ("Yes, . . .I understand . . . Unhuh, . . "). These verbal forms of inserting acknowledgement occur every six seconds in the Consultative style. When the group size has grown too large, the Formal style is employed and,

> " . . .the insertions then may overlap, causing semantic confusion, or each listener must space his insertions out beyond the biological limit of about thirty seconds; either of these results then causes this or that group-member to withdraw by becoming catatonic or absent, or to begin speaking in formal style and thus to render the others catatonic or absent." (Joos, 23:34).

But the speaker employing Formal style also attempts to render himself absent by systematically avoiding any reference to himself as "I, me, mine" etc. using instead "One must . . ." The Formal style, then, seeks to rule out involvement by the speaker as well as the listener, thereby effectively reducing personal access. One has, instead, access to a figure of authority.

The fifth style, the "Frozen," can best be described by using Joos' summary description of the function of each style (Joos, 23: 41-42),

> "Good intimate style fuses two personalities. Good casual style integrates disparate personalities into a social group which is greater than the sum of its parts, for now the personalities complement each other instead of clashing. Good consultative style produces cooperation without the integration, profiting from the lack of it. Good formal style informs the individual separately, so that his future planning may be the more discriminate. Good frozen style, finally, lures him into educating himself, so that he may the more confidently act what role he chooses."

"Frozen" style, of course, is written, and is intended for people "who are to remain social strangers." Its advantage is that it can be reread, re-experienced, and therefore open to reinterpretation.

Linguistic style, then, is the manipulation of linguistic structures which inform interacting individuals of the relative degree of access they are permitted with respect to one another. On the Intimate side of the style scale, one person has total access to another (only feelings from inside the speaker's skin are communicated); but, as a consequence, the addressee must supply much of the information to make the message meaningful. On the Frozen side, all members of a community have potential access to the information of the speaker (author), but they do not have access to the speaker himself! Intimacy and anonymity are curious complements in the orchestration of interpersonal access resulting from the manipulation of structures and the concomitant regulation of boundaries whether microspatial, social or conceptual.

If we accept Altman's (1) generic definition of privacy as involving the ". . .selective control of access to the self or to one's group," then it should be apparent that communality must be considered the functional complement of privacy since it, too, involves the regulation of interpersonal access through the manipulation of structures. Communication, as a process, appears to be the behavioral mediator between the two complements; and should be considered a generic model for man-environment relations (De Long, 8).

We have delineated the role of structure in boundary-regulation and the orchestration of interpersonal access. Structure itself, however, cannot be fully appreciated or identified in behavioral systems unless relationships and contextual logic are considered. It is to these issues that we now turn.

RELATIONSHIPS AND CONTEXTUAL LOGIC

When we say "one is to two as two is to *X*" we are specifying analogues. The *relationships* we are concerned with when dealing with structures and contextual logic are *analogies*. These relationships have several interesting properties: they account for contextual factors, and they result in unitary entities which exhibit invariance. Another way of speaking of these relationships is to say they employ analogue techniques to arrive at digitalization: they transform the continuously variable into the functionally discrete. It is this type of relationship which is central to understanding the relationship between environment and behavior.

What is *a priori* in contextual logic is the *relationship*. It can be considered *a priori* in several senses. First, it may simply be "wired into" the organism as a part of its biogrammar (Tiger & Fox, 30), that is, the relationship can be biobasic (Hall, 19). The synchronization of body movements of the neonate as observed by Condon (5), for example, appears to be "wired in" because the neonate will initially synchronize to verbal organization regardless of the language and its structure, although later in life such microsynchronization of kinesics breaks down when a language foreign to the person is employed. Second, it can be *a priori* in the sense that the culture of the individual *demands* that he effectively incorporate the desired relationship in order to be accepted as a normal, functioning member of the group. In any case, what is requisite to survival for human beings on the cultural level is the recognition of specific sets of relationships.

A critical question must now be raised. Given the viscissitudes of inherent variability both in behavior and circumstances how can relationships which constitute unitary entities and exhibit invariance be expected to be perceived? The answer appears paradoxical: Invariance, or a sense of constancy, is only achieved in the midst of variability. It is the distribution, or *allocation* of variability which leads to perceptual and/or functional invariance.

By way of simple analogy, the fractional relationship "one-half" can be said to be an invariant unit (i.e., it can be transposed, combined and manipulated with other invariant relationships). Furthermore, this invariant relationship maintains its constant identity through the allocation of variability. This can be illustrated in the following hypothetical design problem.

> "Design an environmental context in which it is possible to conclude without contradiction that "1" is functionally equivalent to "2".

To which the solution offered consists of,

Event, "$_1$"
Event, "$_2$"

Environmental elements, "/" and, "2", "4"

arranged as, "/2" and "/4"

and populated by,

"$^1/2$" and "$^2/4$"

since 1/2 = 2/4, it is argued that,

Given the relationship "one-half,"

{½}, consists of /^{1}/ in context, "/2" alternating with /^{2}/ in context "/4"!

The properties of relationships of this type are worth exploring briefly. Invariance is the result of matching different events (numerator) with different contexts (denominator). The relationship {½} can be manifested as "$^1/2$", "$^2/4$", "$^3/6$", or "$^4/8$." Neither the events nor the contexts are constant, both vary continually. *Invariance*, then, can be seen to reside in the *relationship*. We can say that with respect to the relationship {½}, /^{1}/, /^{2}/, /^{3}/ and /^{4}/ are complements, or are complementary with respect to the manifestation of the invariant unit, because we can specify the contexts in which they will occur.

Thus, we can see that *complementarity* leads to *redundancy*, a sense of stability and constancy despite continually changing inputs, through the definition of invariant relationships. The variation of the type we have considered in the manifestation of the relationship {½} can be seen to be *nonfunctional* insofar as it does not alter the relationship perceived.

Another type of variation requires consideration. If "$_1$" and "$_2$" can occur in the same context (e.g., "$^1/3$" and "$^2/3$") or if "$_1$" can occur in different contexts (e.g., "$^1/3$" and "$^1/4$") it is obvious that the relationships *contrast* with one another, and represent differences-which-make-a-difference. The variation present in this case is *functional* and serves the purpose of introducing *information*.

Contextual logic, then, leads to several succinct conclusions regarding man-environment relations. If the numerator of our fractional relationship is considered *behavior* and the denominator is considered the *environment*, or context, we can see that the same behavior in varying environments leads to different relationships, as does different behavior in the same environment.

Or,

$$a_1(C_1) \neq a_1(C_2), \text{ and } a_1(C_1) \neq a_2(C_1)^6 \rightarrow \text{information.}$$

We can also see that if different behavior occurs in different environments we have the *potential* for creating different manifestations of the same relationship. This variation is nonfunctional, introducing redundancy.

Or,

$$a_1(C_1) = a_2(C_2) \rightarrow \text{redundancy.}^7$$

---

6. It is interesting that the first formulation is a definition of magic and illusion while the second formulation is a definition of the scientific experiment. Both yield information. The latter supports Hall's contention (22) that science often avoids dealing with context -- by making it a constant. It is presumably this similarity between magic and science which underlies Levi-Strauss' (24) argument that they are complementary modes of knowledge acquisition (De Long, 11).

7. This formulation differs in character from those preceding, and could be considered a definition of design.

The two types of variation delineated by contextual logic lead to an interesting conclusion concerning the adaptive and evolutionary potential of systems based upon this logic. On the one hand, given the relational unit {½}, if new contexts emerge, say "/10," and "/12," we must create events "5" and "6" in order to maintain invariance. As new contexts arise, then, we continually increase the diversity of events within the system in order to maintain an expression of the invariant unit. Thus, the application of contextual logic to create redundancy leads to its adaptive character. On the other hand, once "6" becomes an event within the system, the possibility exists that we may use it to generate information -- for example, by creating "6/10". Since $a_1(C_1) \neq a_2(C_1)$ ("5/10" ≠ "6/10") we have created a new relationship, {3/5}; and now have an entirely different range for expressing redundancy. The application, then, of contextual logic to yield information gives rise to its evolutionary properties.

Whether we employ contextual logic to create redundancy or information, we can see that there exists an interplay between diversity and unity. Thus,

/Diversity/ ← → {Unity}

Diversity and potential unity are mutual-causal complements: the resolution of diversity to achieve stability, or redundancy, results in unity the search for which leads to further diversity over time. This constitutes a property of some significance as Maruyama (26) has shown. Philosophically, it is an essential point for the design of man's environment, and his future; but one typically overlooked in this culture (Greenbie, 18). Contextual logic leads to invariant relationships based upon the principle of complementarity (De Long, 8); and which function as units obtaining functional significance and stability through contrast.

CONCLUSION: THE PERVASIVENESS OF CONTEXTUAL LOGIC

In this paper we have dealt with a variety of concepts and illustrations which we feel are central and requisite for the study and eventual understanding of privacy and communality. In this conclusion, we will summarize those concepts and deal with their integration.

One of the most fundamental concerns in organism-environment relationships is how, given the intrinsic variability in both the environmental circumstances and the emitted behavior, organisms can maintain a sense of constancy and stability (De Long, 8). What is perceived to be stable and constant is not the environment nor the behavior of an organism; but rather the *relationship* between the two. The sense of stability is arrived at through the operation of a contextual logic which is relational in orientation; and assigns perceptual and functional value to behavior within the context in which it occurs. The relationships to which an organism attends can be considered *a priori*, either because they constitute a portion of the biogrammar or because they are deemed necessary for social purposes by the group into which the organism is to be enculturated. Such relationships are characterized by invariance; and hence can be manipulated either sequentially or configurationally[8] as structural units.

In behavior *systems* which have a communication function, the structure of the system (how behavior is organized) is the prime source of regulation between organisms. Territorial systems, for example, are principally responsible for organizing the spatial relationships between organism in a variety of species. Dominance hierarchies constitute systems for regulating social relationships between members of a group. Kinesic systems serve to regulate the organization of interpersonal relationships within the group while language serves to regulate the conceptual aspects of interpersonal and group transactions. Further, the structures between the different levels of complexity (the spatial, the social and the conceptual) exhibit redundancy, insofar as what is regulated on one level tends to be reinforced on another in a similar manner. The study of dominance-territorial relationships in a small group, for example, indicated that changes in the spatial structuring of the group were congruent and concurrent with changes in the hierarchical structuring of the group and with the conceptual structuring as well (De Long, 6, 7). In other words, messages at one level of organization are consistent with messages from other levels at the same time. In fact, when spatial social and conceptual structures are manipulated to yield different, contradictory messages, people are placed in what Bateson (3) referred to as a "double-bind" -- a situation in which stability may be jeopardized.

The relative economy and productivity of systems whose structures are based upon contextual logic is a feature of considerable significance. The same event can function different ways within the structure depending upon its contextual environment. This can be seen in the final illustrations in Figure 5. Using two different events (figures) and two different context (grounds) we can see in the first instance that $a_1 \neq a_2$ when both occur in the same context. In the second instance we can see that $a_1 \neq a_1$ when it occurs in

---

8. Sequence is here interpreted to consist of configurations-subject-to-temporal-organization. Language, of course, is sequential. The microspatial environment includes both configuration and sequence.

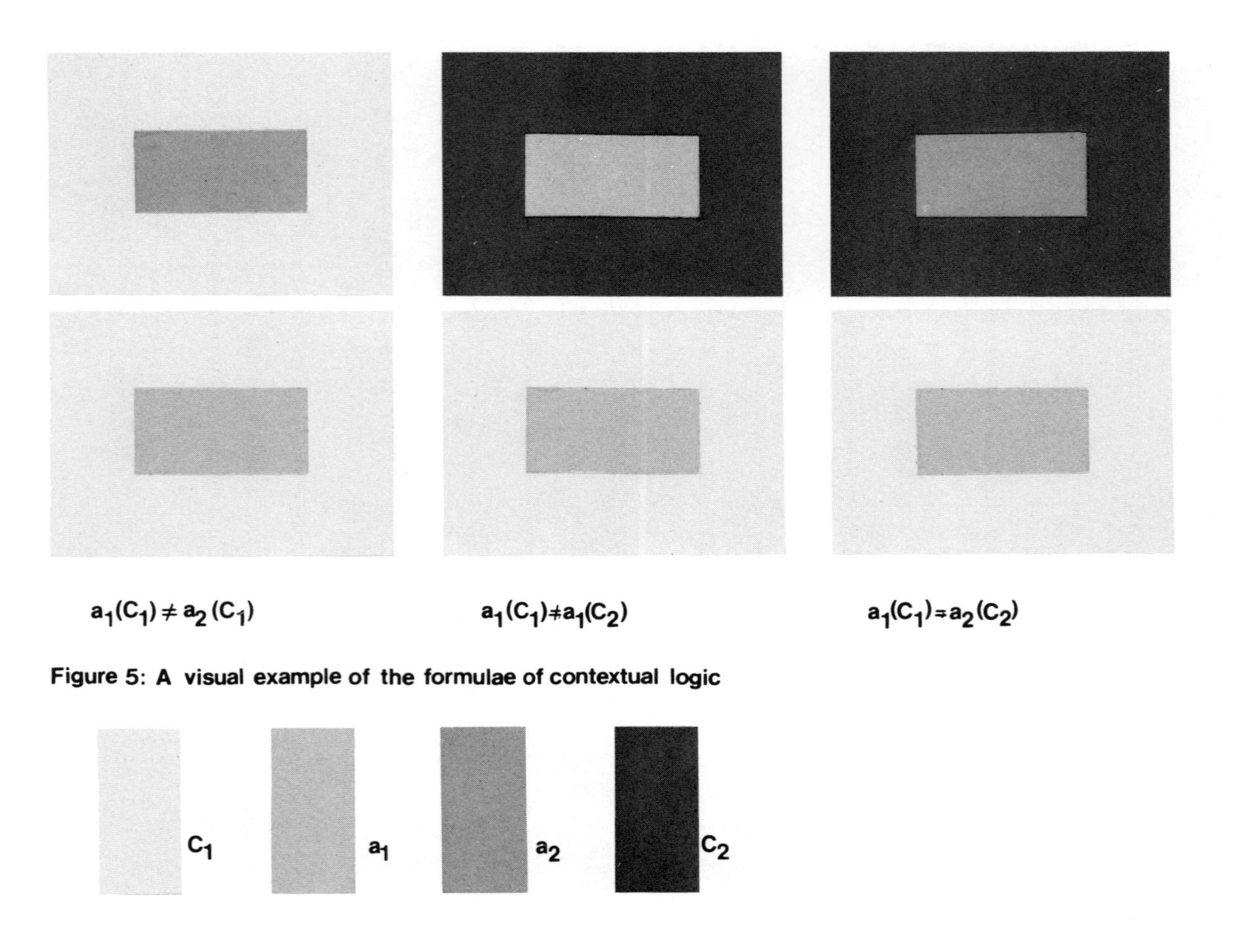

Figure 5: A visual example of the formulae of contextual logic

in two different contexts. The perceptual and functional value of the same event is altered by the context in which it occurs. Finally, we find that $a_1 = a_2$ when they occur in complementary contexts.

Similarly, the distance-variants we discussed earlier in Tables 1 and 2 exhibit a curious logical property. Namely, a = b in the Personal-Far relationship, while b = c in the Social-Close relationship. *But nowhere does a = c.* And additionally, even though a and b are equivalent within the Personal-Far relationship, we see that the difference between a and b serves to distinguish the Personal-Far relationship from the Social-Close relationship for the side-by-side orientation. Thus, the distance-variants a and b are being employed by the system for their similarity as well as their differences. Analogous situations can be found in Language as well. In spoken English, for example, {s} and {z} function to differentiate otherwise identical forms -- "sing" and "zing" as well as "bus" and "buzz". The differences between the two sounds are functional; and we can see that $a_1 \neq a_2$ when they occur in the same context. At a different level of linguistic structure, however, we find that /-s/ and /-z/ are treated as being equivalent because of their similarities. In making plurals we find that these sounds constitute two of the variants; and that /-s/ is added to forms ending in voiceless consonants ("cat" + s), while /-z/ is added to forms ending in voiced consonants ("dog" + z). Here we see that $a_1 = a_2$ because each occurs in a complementary context. We can also see that a_1 in one context has a different value than it does in another (i.e., $a_1 \neq a_1$). The same /-s/ and /-z/ are also employed in making possessives, having the same complementary distribution. The question then arises as to how the speaker knows whether /-s/ is a plural or a possessive? The answer, of course, is context.

Thus, statements such as,

$$a = b = c, \text{ but } a \neq c; \text{ and } a \neq a$$

are not contradictory at all when the contextual aspects are considered, because the perceptual and/or functional value of any given event is always a function of its context.

The role of complementarity in contextual logic applies with equal importance to systems at a wide variety of levels and underlies the structure of visual, auditory and conceptual systems. It is a cogent to the structure of DNA and Language as it is to the *I Ching* (Stent,29). Indeed, complementary appears as central to our comprehension of social behavior as it has been in physics and as it was in uncovering the structure of DNA. With regard to sociality as the overall concern, it would seem that communality and privacy function very much like the patches of gray in our previous illustration. It is obvious to us that the two states appear to function differently when we hold the context constant. But two-thirds of the problem still awaits inquiry -- delineating the various functional values of each state as well as identifying the genuine, complementary nature of communality and privacy. This will require the employment of contextual logic. The result should be the beginning of a clear comprehension of the adaptive and evolutionary potential of human sociality.

REFERENCES

1. Altman, I. Privacy: A conceptual analysis. *Environment and Behavior*. 1976, *8:* 7-29.

2. Anderson, B. F. *Cognitive Psychology*, New York: Academic Press, 1975.

3. Bateson, G. *Steps to an ecology of mind*. New York: Ballantine, 1973.

4. Bernstein, B. *Class, Codes and Control*, Vol. I, London: Routledge & Kegan Paul, 1971.

5. Condon, W. S. & Sander, L. W. Neorate movement is synchronized with adult speech: Interactional participation and language acquisition. *Science*. 1974, *183:* 99-101.

6. De Long, A. J. Dominance-territorial relations in a small group. *Environment and Behavior*, 1970, *2:* 170-191.

7. De Long, A. J. Dominance-territorial criteria and small group structure. *Comparative Group Studies*. 1971, *2:* 235-266.

8. De Long, A. J. The communication process: A generic model for man-environment relations. *Man-Environment Systems*, 1972, *2* (5): 263-313.

9. De Long, A. J. Aspectual and hierarchical characteristics of environmental codes. In: W. F. E. Preiser (Ed.) *Environmental Design Research: Selected Papers*, Vol. I. Stroudsburg, Pa: Dowden, Hutchinson & Ross, 1973.

10. De Long, A. J. Kinesic signals at utterance boundaries in preschool children. *Semiotica*, 1974, *11:* 43-73.

11. De Long, A. J. The scientist and the sorcerer: Creating man-environment systems. *Architectural Student*. 1974, *23:* 16-18.

12. De Long, A. J. The use of scale-models in spatial-behavior research. *Man-Environment Systems,* 1976, *6:* 179-182.

13. De Long, A. J. The accuracy of spatial perception by informants in scale-model environments. *Man-Environment Systems,* 1977, *7:* 55-58.

14. De Long, A. J. Proxemics and context: An empirical analysis. 1977, Ms.

15. De Long, A. J. Yielding the floor: The kinesic signals. *Journal of Communication.* 1977, *27:* 98-103.

16. Esser, A. H. Theoretical and empirical issues with regard to privacy, territoriality, personal space and crowding. *Environment and Behavior,* 1976, *8:* 117-124.

17. Gleason, H. A. *An introduction to descriptive linguistics.* New York: Holt, Rinehart and Winston, 1955.

18. Greenbie, B. Design for Diversity. Amsterdam: Elsevier Scientific Publishing Company, 1976.

19. Hall, E. T. *The silent language.* Garden City, N.Y.: Doubleday & Co., 1959.

20. Hall, E. T. A system for the notation of proxemic behavior. *American Anthropologist,* 1963, *65:* 1003-1026.

21. Hall, E. T. *The hidden dimension.* Garden City, N.Y.: Doubleday & Co., 1966.

22. Hall, E. T. *Beyond Culture.* Garden City, New York: Doubleday & Co., 1976.

23. Joos, M. *The Five Clocks.* New York: Harcourt, Brace & World, 1967.

24. Levi-Strauss, C. *The Savage Mind.* Chicago: University of Chicago Press, 1966.

25. MacLean, P. On the evolution of three mentalities. *Man-Environment Systems.* 1975, *5:* 213-224.

26. Maruyama, M. The second cybernetics: deviation-amplifying mutual casual processes. *General Systems,* 1963, *8:* 233-241.

27. Scheflen, A. The significance of posture in communication. *Psychiatry,* 1964, *27:* 316-331.

28. Sommer, R. *Personal Space.* Englewood Cliffs, N.J.: Prentice-Hall, 1969.

29. Stent, G. S. *The coming of the golden age: A view of the end of progress.* Garden City, N.Y.: Doubleday & Co., 1969.

30. Tiger, L. & Fox, R. *The imperial animal.* New York: Dell, 1974.

31. Von Foerster, H. The logical structure of the environment and its internal representation. *Man-Environment Systems.* 1974, *4:* 161-170.

PRIVACY REGULATION IN PUBLIC BATHROOMS: VERBAL AND NONVERBAL BEHAVIOUR

Michael G. Efran and Charles S. Baran

Division of Life Sciences
University of Toronto at Scarborough College
West Hill, Ontario, Canada

There are two ways to study man-environment relations. One is to bring the environment into the laboratory and the other is to bring the laboratory out into the environment. On the principle that it is easier to bring Mohammed to the mountain than it is the mountain to Mohammed, the later alternative was taken and the studies reported here were conducted within a series of public mens rooms located in a large university building. The building also houses classrooms, offices and other laboratories. This seemingly unlikely site was selected because the studies deal with the behavioural regulation of privacy and it seemed that, at least in North America, this location guaranteed privacy needs of an intense nature (though just why this should be so is somewhat of a mystery).

The studies reported here were conducted to document a pattern of behavior well known to most men on this continent. The rule which describes the spatial component of this behaviour was simply stated by Robert Sommer after he had made some casual observations in a mens room which at the time was frequented by a group of conventioning psychologists.[1] The rule is that one should, whenever possible, leave at least one vacant urinal between yourself and anyone else present. Accompanying the resulting spatial isolation is a familiar pattern of gaze avoidance and withdrawal. In the first study reported here, an automatic recording device was used to determine the extent to which this distancing rule is actually followed. The second study was conducted to document the other withdrawal mechanisms which occur when the taboo on interaction is broken. In both studies comparable behaviour was recorded while subjects washed their hands at the washbasins since this activity,

---

[1] Personal Communication, February 25, 1971.

though involving the same basic spatial parameters, is generally not associated with the heightened privacy needs which accompany elimination.

STUDY I

A pressure sensitive switching mat sandwiched between two sheets of black rubber floormat and placed beneath the sinks and urinals was used to activate the pens of a concealed battery driven event recorder whenever anyone stood in front of these units. This provided a complete (and completely anonymous) record of when each sink or urinal was used. The apparatus was installed, on different days under either three linearly arranged sinks or three linearly arranged urinals. Three days use of each type of unit was sampled on a random basis (i.e., the day was selected randomly and the entire day's activity at one type of unit was recorded).

The results were perfectly clear. A distancing rule such as that described by Sommer operates at the urinals but not at the sinks. Seventy-six men chose the nonadjacent urinal and 17 chose the adjacent one ($z = 6.0$, $p<10^{-6}$). Fewer men washed their hands and the test situation only arose 17 times at the washbasins. However, no distancing pattern was evident with 8 men selecting the nonadjacent washbasin and 9 the adjacent one. These patterns appear to completely override preferences displayed by solitary users for the urinal nearest the door and the sink nearest the towel dispenser.

STUDY II

In this study a student confederate entered the bathroom with the intention of violating the rule to avoid even minimal social interaction in the mens room. First he entered the bathroom and determined whether it was being used by a single other man who was either washing his hands or urinating. If this condition was met the man was tagged as a subject and the confederate proceeded to use the facility directly adjacent to that already in use. Once engaged in the appropriate activity the confederate initiated interaction by turning to the other man and saying, "Wow, is this place ever a drag!" The man's first verbal response (if any) was memorized and recorded verbatim after the confederate left the bathroom. In addition, the confederate made notes on the nonverbal elements of the interaction which he had initiated through his verbalization.[2] Fifteen men were approached at the urinals and 15 others at the washbasins.

---

[2] The confederate had been trained in the observation of nonverbal behaviour for a previously conducted study.

The results of this study were quite consistent with those of the first study and were also very clear. At the washbasins men generally responded to the verbal probe in complete sentences. They looked at the confederate as they spoke and appeared relaxed and friendly. At the urinals they hardly spoke at all. They looked down or away from the confederate (even when they spoke to him) and appeared uneasy. The most common verbal response at the urinals was either complete silence or a one word paralinguistic utterance such as "mmmh." Mouth gestures such as lip biting and mouth corner retraction as well as fidgeting and postural rigidity were observed at the urinals but not at the sinks. At the sinks the subjects' verbal responses average 10.7 words while at the urinals they spoke a mean of only 1.6 words (t = 7.4, df = 28, p<.001).

DISCUSSION

Sommer has documented the way people use space and body orientation to maximize isolation in a public library (15) and Patterson, et al. (14) have shown that when intrusions do occur the person who is intruded upon performs characteristic acts which appear to insulate or distance him from the intruder. These acts consist of things such as reorienting the body or propping a hand to the face on the side of the intruder. Dabbs has found that similar postural shifts and away-movements are prompted when confederates stand too close to pedestrians on a city sidewalk (6) and Efran and Cheyne as well as Knowles have shown the lengths to which pedestrians will go to avoid intruding on the shared space of others (5, 7, 12). Ellsworth, using a prolonged stare as a means of intrusion, found that both pedestrians and drivers responded to this violation of their privacy with accelerated departure speeds (8).

From one perspective (16) these nonverbal acts and the sequences in which they occur *are* privacy regulation. That is they are the behavioural mechanisms of privacy regulation and they may or may not be accompanied by compatible cognitions and verbal statements. Although some of the same external variables may be instrumental in establishing both the nonverbal regulatory behaviour and the conceptualizations which accompany them the primary locus of control of the latter is conceptualized within a socio-cultural framework while the former are attributed to a bio-social sphere whose primary products are feelings and actions rather than words and thoughts. The approach taken here is that in order to completely understand the nature of privacy needs it is essential to learn more about these nonverbal mechanisms; their significance and the conditions which elicit them.

Although we often think of privacy needs in terms of longer time intervals (e.g. an hour in the library or a weekend alone) contemporary urban life enforces numerous interactions of a more momentary nature such as those examined here. Our view is that although the consequences of these interactions do not ordinarily filter into our personal awareness of our needs they are nonetheless essential components of social regulation and the pattern of nonverbal behaviour which customarily emerges has a great deal to do with our general state of mental and physical well-being. Calhoun has shown, with mice, the intimate connection between population dynamics, behavioural social regulation and physical well-being (3). Calhoun's findings emphasize the importance of simple acts which animals use to signal and engage in things such as aggression and affiliation. Although the links between these behaviours in mice and other animals and equivalent acts in human beings are not as firmly established as we would like to see, our view is that there are sufficient empirical grounds to warrant further study of the way that things such as arousal reduction through gaze avoidance (e.g., 2, 4, 13) function in naturally occuring situations and are related to more anthropomorphic concepts such as privacy.

In the two studies reported here it is clear that men engaged in avoidance behaviour and appeared uneasy at the urinals but not at the sinks. This is an interesting contrast. Many men, when presented with these findings, reason that it is a mild fear of homosexual encounters which prompts their own behaviour at the urinals. Acceptance of this view, which may afterall be correct, strengthens the argument that the observed behaviours are indeed directly related to privacy regulation. However, one can also ask whether the incidence of homosexual encounters actually initiated at urinals doesn't make this widespread fear (if it exists) a form of cultural paranoia. If the fear is not widespread where does the behaviour originate? Psychoanalytic thought, though discredited in some circles, deals extensively with the roots of irrational and half realized fears. It is interesting to note that this mode of thought emphasizes the developmental role of both eliminative acts and sexuality and their link to the core of adult personality. Is it possible that these linkages are also operative in association with the behaviour observed here and that men exhibit greater defensiveness at the urinals than at the sinks because, so to speak, of the nearness to hand of sexual cues? Is it possible that the association which seems so deeply entrenched in this culture between heightened privacy needs and what Kira (11) has termed the sexual/eliminative amalgam has a more fundamental causation underlying the cultural taboo? Is it also possible that some of the same factors and half voiced conceptual links which are relatively easy to identify in the bathroom are also operative in other situations where people display similar behaviour (e.g., at lunchcounters and on park benches)?

By itself the behaviour recorded in these studies is probably not very important. No one has ever been known to have suffered a fatal heart attack because someone else stood too close to him in the bathroom. However, research has shown that distancing, eye contact and other behaviours do appear to commonly function as part of a complex system of nonverbal regulation which keeps arousal at a comfortable or adaptive level (e.g., 1, 9). People seem to use behaviours similar to those observed here when they gaze avoid on subways and in elevators, when they distance themselves from others on park benches and in waiting rooms and when they avoid others' pathways on city streets. Although we rarely focus our verbal awareness on these behaviours they appear to be instrumental in the regulation of a good deal of daily activity and what we learn about their significance in one situation may permit inferences about their significance in other situations. Collectively they may represent a sizeable portion of the hustle and bustle which is generally acknowledged to accompany urban life. Attention to them may tell a good deal about the way people experience others in many different environments and attention to them in design, collectively speaking, might well have some impact on the way built environments are experienced. For example, in total institutions where depersonalization is often a very serious problem and where these systems of nonverbal regulation are already disrupted or heavily taxed, (e.g., 10) attention to details such as those of washroom design may in fact noticeably influence the quality of life experienced in these places. Splash boards which extend beyond eye level, for instance, might provide a simple and economical way of ensuring the isolation which is otherwise sought behaviourally.

Perhaps an anecdote taken from my childhood will serve to illustrate that the observations documented in these studies are both important and not equally obvious to everyone. I grew up in a large American city and attended a large, impersonal and somewhat imposing elementary school. As was then common practice the doors to all of the toilet cubicles had been removed. Informal conversation with my peers today has shown that I was not the only child who let social pressures outweigh internal forces. I was not the only child who therefore postponed, sometimes with considerable difficulty, gratification of an imperative biological need until the act could be accomplished less publicly at home. Obvious? Perhaps, but someone ordered the doors removed!

REFERENCES

1. Argyle, M. *Social interaction.* New York: Atherton, 1969.

2. Argyle, M. and Dean, J. Eye-contact, distance, and affiliation. *Sociometry,* 1965, *29*:289-304.

3. Calhoun, J. B. Space and the strategy of life. In A. H. Esser (Ed.) *Behavior and environment.* New York: Plenum, 1971.

4. Chance, M. R. A. An interpretation of some agonistic postures: The role of "cut-off" acts and postures. *Symposia of the Zoological Society of London*, 1962, *8*:71-89.

5. Cheyne, J. A. and Efran, M. G. The effect of spatial and interpersonal variables on the invasion of group controlled territories. *Sociometry*, 1972, *35*:477-489.

6. Dabbs, J. Sex, setting and reaction to crowding on sidewalks. *Proceedings of the 80th Annual Convention of the American Psychological Association*, 1972, 205-206.

7. Efran, M. G. and Cheyne, J. A. Shared space: The cooperative control of spatial areas by two interacting individuals. *Canadian Journal of Behavioural Science*, 1973, *5*:201-210.

8. Ellsworth, P. C., Carlsmith, J. M., and Henson, A. The stare as a stimulus to flight in human subjects: A series of field experiments. *Journal of Personality and Social Psychology*, 1972, *21*:302-311.

9. Exline, R. V. Visual interaction: The glances of power and preference. In J. K. Cole (Ed.) *The Nebraska Symposium on Motivation*, 1971, *19*:163-206.

10. Grant, E. C. Non-verbal communication in the mentally ill. In R. A. Hinde (Ed.) *Non-verbal communication.* Cambridge: University Press, 1972.

11. Kira, A. *The bathroom: Criteria for design.* Ithaca, N. Y.: Center for Environmental Studies, Cornell University, 1966.

12. Knowles, E. S. Boundaries around group interaction: The effect of group size and member status on boundary permeability. *Journal of Personality and Social Psychology*, 1973, *26*:327-331.

13. Nichols, K. A. & Champness, B. G. Eye gaze and the GSR. *Journal of Experimental Social Psychology*, 1971, *7*: 623-626.

14. Patterson, M. L., Mullens, S. and Romano, J. Compensatory reactions to spatial intrusions. *Sociometry*, 1971, *34*:114-121.

15. Sommer, R. The ecology of privacy. *Library Quarterly*, 1966, *36*:234-248.

16. Tibbetts, P. and Esser, A. H. Transactional structures in man-environment relations. *Man-Environment Systems*, 1973, *3*:441-468.

Section 4: Residential Design

INTRODUCTORY NOTES

Of all behavioral environments, undoubtedly the one in which privacy constitutes the strongest imperative is housing. The home, by definition, is a private place. To say that it is a private place is, of course, to say that it is *not* a public, or communal, place, and we return to the theme with which this book opened and which has been reasserted throughout the discussions so far. In other words, we cannot look at residential design as a source and focus for privacy, except by examining the boundaries which separate a particular residence from the community in which it is set. Even the hermit, seeking solitude in the wilderness, affirms his inexorable relationship to a larger urban environment by placing himself as far as possible from it. Most people do not wish to isolate themselves from communality with other human beings and could not achieve such isolation if they did wish to. But what is significant for the designers of residential environments is that the extent and nature of desired involvement of the residents of one unit with those of others varies greatly, not only with culture and individual personality, but with life-cycle and a wide range of other circumstances. Thus residential design is the ultimate of boundary regulation.

This section deals empirically with such regulation. In the first paper, Beck and Teasdale examine the dimensions of boundary regulation in a cross-cultural study of French-Canadian and English-Canadian families living in multiple dwelling units. The authors used an interview rather than the often abused questionnaire technique to probe the residents' feelings about such aspects of living as security, leisure and maintenance. This study is somewhat unique in this volume and in man-environment research in general, in that it involves the active collaboration of a social scientist with architects in a study directed primarily not to general knowledge of behavior, but toward specific design criteria. The authors stress this point and profess what they preach by arriving at a set of specific "Site Social Requirements" which, in the technical language of the times, might be called SSR. Such specificity requires considerable courage in a responsible social scientist, whose data are never conclusive enough for his own satisfaction and who feels much safer if he can limit recommendations to a call for further research. These authors do that too, but their practical criteria will be welcome to designers who read this volume for the purpose of getting help in designing better buildings. The recommendations are summarized in the main paper, but, since they were funded separately, they are restated in more detail as an Appendix, which the reader is urged not to overlook.

Appropriately, Beck and Teasdale start out by underscoring the basic concept developed in particular by DeLong in Section 3, that the nature of privacy, along with other social relationships, can only be considered in specific environmental and functional contexts. Their study shows that privacy needs vary greatly, not only in different cultures (here French and English Canadian), within the five different designed environments studied, but also in time and place according to the various activities of the families involved. They note that a particular family may be sensitive to visual privacy while eating, but not while gardening, for example. And yet, despite the complexity and fluidity of the concept, privacy remains a recognizable common variable across the board in all groups.

By accepting relationships, rather than *a priori* absolutes, as the basis for the research design, the authors have empirically demonstrated the "contextual logic" advocated theoretically by DeLong. Privacy, then, is both a variable and a constant, depending on context, and so, necessarily, is communality. The designer's first problem is to determine the relevant context. Using DeLong's analogy, we may conclude that if the need for privacy is taken as a numerator, both the culture of the subjects and the social and functional environments may be taken as denominators. Conversely, taking any one of the designed residential environments as a constant, or any one of the cultures as a constant, privacy needs will vary with activities and circumstances. These may involve not only physical settings and social or functional events, but concepts, such as the definition of "a good neighbor." It is a measure of the effectiveness of this approach that, despite the complexities and the multiplicity of variables considered, Beck and Teasdale could indeed come up with specific criteria for design.

In the second paper of this section, Jaanus and Nieuwenhuijse address a somewhat similar problem from the opposite direction. In this empirical study of predictors of housing preference in a Dutch suburb, they are dealing with a population that is culturally homogeneous (at least as far as the data show). They looked at those variables in the design of several multi-family housing areas which correlate with expressed preference to live in one or the other. Using statistical analytical methods, they collapsed 36 different characteristics of these developments into three basic dimensions. These in effect represent basic sets of relationships, or physical contexts. One of them, called by the authors "comsit," pertains to those perceived physical factors which were subjectively taken to provide comfort and a situation leading to privacy. The most prominent of these is the latter, the degree to which the physical layout appeared to offer privacy. As we might expect from other discussions in this volume, this correlated highly with housing preference. The other two dimensions they worked with did not, but analysis of one of these negatively correlating factors is highly instructive.

This negatively correlating factor is one the authors call "social functions." The residential areas they studied had grown up successively beyond an older center, which was the only one of the areas to contain a shopping center and lively but heavily trafficked streets with various urban attractions. While this area rated very high in the list of social functions considered, it correlated poorly with preference for the area as a place to live. That is, people seemed to like it as an activity center to go to, but not to live in. The contextual consideration of such variables obviously can avoid the intellectual polarizations of many designers and planners who deplore the "sterility" of suburbia and extol lively downtowns as an ultimate good. Nieuwenhuijse and Jaanus suggest that, at least for some middle-class people, dormitory suburbs should not be considered *a priori* "bad," as so many social theorists do, and the fact that many people equate the lack of urban communal activity with privacy does not suggest that in another context these same people may not relish and seek out high density, high intensity commercial downtown centers. On the other level, as these authors note, a different population might have responded quite differently.

The third factor which failed to correlate with preference for living areas is one that also has been an *a priori* "good" among designers and planners--that of open space, here called "green space." There were a number of considerations which may have affected this variable as a poor predictor of housing preference, but one point is worth noting. The fact that in this case green space was not a determinant of preference does not suggest that green space in itself might not be desired by these people or that, all other things being equal, it might not contribute ultimately to housing preference. But from the point of view of designers, it is important to consider that publicly accessible green space may reduce or increase the sense of privacy, depending on how it is structured, in other words, on the design context. If green space is designed to afford privacy it might indeed become a predictor of preference.

DIMENSIONS OF SOCIAL LIFE STYLE IN MULTIPLE DWELLING HOUSING[1]

Robert J. Beck and Pierre Teasdale

Centre de recherches et d'innovation urbaines
Université de Montréal
Montréal, H3T 1T2

This paper is an inquiry into selected dimensions of the social lifestyle of 75 French-Canadian and English-Canadian families inhabiting multiple dwelling housing in 4 cities in Eastern Canada. Our purpose was to investigate the way in which the physical environment was implicated in the initiation, development and sustenance of several kinds of social lifestyle variables including activities (e.g., visits), relationships (e.g., friendships), values (e.g., privacy attitudes toward neighbors), community image (perceived solidarity), personality and several others. The study examines such behavior with respect to cultural and family demographic variations in the sample which consisted uniquely of whole families with 2-3 young children. An in-depth methodology consisting of a 115-question interview, furniture maps, field notes, observation and photographs was used with the families to generate criteria for government standards for user program requirements in Canadian housing. Distinct cultural and family type differences and preferences are noted which have implications for the design of multiple dwelling housing. The work also sheds light on the empirical study of privacy and its relationship to other dimensions of social life. It is concluded that privacy is best understood when studied in a particular genre of environment (housing) in the context of other social environmental behaviors in relation to other areas of housing lifestyle; and by using design research methodology (user analysis) in preference to more disciplinary approaches.

---

[1] The authors gratefully acknowledge the support of this research by the Central Mortgage and Housing Corporation, Government of Canada.

OVERVIEW

We are now witnessing a strong rise of multi-disciplinary activity in the analysis of privacy concepts and meanings and reviews of empirical studies focusing in whole or in part on privacy relations. In this paper, we intend to analyze privacy in the context of a range of social relations in housing. We do not expect that privacy or other social manifestations, standards and meanings remain constant across environments. We present a body of data collected for the Central Mortgage and Housing Corporation of Canada, who mandated an interdisciplinary team of architects and an environmental psychologist to determine social needs and spatial requirements for residents of moderate income inhabiting low-rise low-density multiple dwelling housing. The data show that the quality and quantity of social relations, including privacy, varies distinctively among each of 5 projects studied in 3 provinces of Eastern Canada, and that in particular there are strong cultural (French-Canadian/English-Canadian) differences in such relations.

In our opinion recent speculation about the categorization of concepts within the privacy molecule (1, 6) and investigations into the meaning of the construct (8) is extremely valuable in that such work brings both theoretical and developmental perspectives on the concept but, in the end, the approach is more traditionally psychological than a man-environment approach. To aid designers the problem should be studied as it operates in a live, finite natural environmental social system. One kind of system (e.g., housing) cannot be completely generalized to another (e.g., hospital).

Our purpose is to extend design oriented research (cf. 5) into social relations in specific environments as a means of defining and coping with the problems of privacy. Design oriented approaches have traditionally sought to uncover user *needs* and focused on human *activities* in space as the primary units of categorization upon which to model behavioral satisfaction with specific kinds of environments or to generate descriptions of potential or ideal environments desired by users. It is our finding that privacy, among other environmental social relations, varies according to what kind of activity we are talking about. Thus food-related activities have their own standards of privacy and community relations when compared with leisure activities. Each activity system or circuit (2, 3, 7) has a unique set of acts and situations and involves different definitions, meanings and standards of privacy.

Thus a family may be sensitive to being overlooked (visual privacy problem) while eating, but not while working on the garden. Even within a class of activities like leisure activities one may be sensitive to privacy while playing cards but not when playing ball. Clearly privacy standards vary with the activity type, the environment, and momentary contexts. The point we are making is that privacy is much too situational and variable a dimension of social relations to try and propose overall categorization schemes describing the sense of privacy in general even for a delimited class of people. Yet we admit that although privacy has this fluid character, we know that in certain individuals or families the sense of privacy is quite distinct and a privacy style appears across the board, in both activities and environment. On this end, privacy attitudes and values have hardened into character and become a more permanent part of the personality structure. In the present work we are finding it particularly difficult to disentangle from both cultural and demographic mediating variables.

Now it is very tempting for psychology in particular and social scientists in general to focus on the place of privacy in a new social psychology or in social relations. After all, it is exciting to discover a new relatively unresearched part of the personality molecule. We see in this work a certain danger that the mandate of the man-environment systems analyst and environmental psychologist is not being satisfied. For us the mandate, stated simply, is to generate information which may be applied to the understanding, evaluation, and design or redesign of environments. Along the way, we most certainly want to understand people's use, feelings, values, cognitions, etc. in the context of environments. The danger is that environmental psychology may be settling back into intradisciplinary activity when what is called for is more research in conjunction with design professionals.

SOCIAL LIFESTYLE

In our definition, social lifestyle in housing consists of a set of social activities and interpersonal relations which are regulated by social values, personality, culture and environmental design, with varying degrees of commitment and intensity. The relations can be minimal (nods, exhanges of greetings, bits of news, etc.) or in-depth (mutual visits among residents designating each other as friends). Other activities are functional expressions of social relations and are less "sociable". We refer here to acts which involve the exchange of products and services - borrowing and lending, what might also be described as consumer or economic behavior in a social context, or social-economic behavior if that were somehow not misleading. Now privacy is a value which is included in both social or sociable and functional social behavior since the private person or family will probably be less sociable and more vulnerable to feelings of invasion if a neighbor comes borrowing. Friendships which emerge among neighbors may come out of sociability and/or consumer relations and become friendships. Where the resident is privacy oriented, there will be fewer relationships designated as friendships and strong attitudes may be expressed that one does not want to have friendships or even minimal contacts in a housing project.

In our study, we researched the following variables which were hypothesized to comprise dimensions or useful means of modelling social life:

1. Social activities: such as visiting, entertaining, borrowing and lending, concurrent leisure activities;

2. Social values: such as what makes for a good neighbor, both in terms of demographic characteristics and day to day behavior;

3. Social relationships: such as the number of friendships, and frequency of minimal contacts;

4. Community image: which attempts to determine the degree of community solidarity within the housing projects;

5. Social cognition: such as extent of knowledge of neighbors' family characteristics (name, number of children, parents' occupation);

6. Social development: in which socialization mechanisms (such as formal introductions and informal introductions via children can be traced);

7. Social-physical environment: in which location and physical equipment were studied to determine to what extent they encouraged or led to privacy or social interaction.

There is considerable overlap in these dimensions with the work of Carey and Mapes (5) which we admire greatly, although their instruments focussed more directly on the development of a sociology of relations among residents in privately owned detached or semi-detached housing over a long term (1 year) period, while our research used more contact at a single point in time among rental residents in multiple dwelling housing with a stronger sense of physical community. We had to reconstruct social development from the residents' memories. Lynnette Carey and Roy Mapes, who are sociologists, naturally came to the conclusion that probably demographic factors are more important in governing social relations than spatial factors in determining the quantity, range and qualitative dimensions of social life, but that ultimately it is the distribution of family types across the site that is paramount in explaining why social relations developed (5).

In this latter finding we heartily concur and come to the same recommendation, to wit that the planning of multiple dwelling housing or estates must include both family type and locational criteria. We also come to similar conclusions concerning the developmental mechanisms, most notably the role of children, introductions, work status of the wife, Tupperware parties, self-introducing residents and the presence of relatives. The English study has the advantage of being able to examine the early development of the relations whereas we come on the scene anywhere from 6 months to 5 years after residents have moved in. They show that acts of helping neighbors within a short time after moving in predict a stronger index of social bonds later on.

Another complicating factor was the presence of community equipment in our physical sample. We shall see that such equipment is a powerful machinery for initiating and sustaining more casual, informal leisure oriented social relations. In this case some of the social energy available may become siphoned off into community relations or the absence of equipment apparently is positively related to a high intensity of proximate ("next-door type") neighboring.

SOCIAL LIFE IN HOUSING

By social life in housing we refer to the pattern of social activities which take place anywhere in the houses or on the site of housing defined as a housing project, that is, multiple housing of a characteristic type.

We have assigned to these social relations 4 distinct scales and definitions of space:

1. The family space within the housing unit.

2. The family space immediately around the housing unit, including characteristic front and rear outdoor space (patio, front yard, etc.)

3. The neighboring space which includes blocks of attached houses and natural groupings of houses.

4. The community space which includes community equipment and common space.

The present study is largely concerned with social relations in 3 and 4, above, and a forthcoming report will be issued on social relations in the house and house adjacent space. Thus the social relations to be examined primarily are those which take place among proximate neighbors and among communities or interest groups sharing and co-using community space.

It must be stated that in housing of this type there is a heightened sense of the housing as an integral unit because the project does not blend into the urban landscape and therefore it does not have the multi-definitional character that may be found in neighborhoods. So, in this housing there is a relatively self-contained system with its own microcosm of social life and the pattern of interaction for the most part is much more among members of the project. The sample projects are also a form of specialized housing with respect to family type, range of children, stage of the family mobility cycle and income. Such projects therefore provide design researchers with relatively defined and finite social systems to investigate.

LIFESTYLE AND SOCIAL LIFESTYLE

The study of social life was the primary activity system studied. However, our object was also to evaluate housing in relation to other important classes of behaviors taking place in and with respect to the environment:

Maintenance: activities which are performed in the service of maintaining any part of the project, whether by resident or management.

Security: efforts to reduce insecurity which is experienced by residents or management and activities which are performed to increase security, including activities such as installing locks, surveying children.

Child: expression of feelings and values which are held with respect to children and activities performed by children, and by adults on behalf of children. Includes activities from all other activity groups, e.g. maintenance, security, social.

Leisure: activities which are performed by children and adults in leisure time and/or for recreational purposes. Includes activities such as sports, games, outdoor life, gardening, and indoor activities.

Food: activities which are performed in relation to the preparing, cooking, serving and maintenance of foods and related equipment.

We wish to establish the importance and unique character of these behaviors in user design programming. Satisfaction of the objectives of such behaviors was found to be highly implicated in overall housing satisfaction. Since the categories are useful for organizing design program information it behooves us to ask how such behaviors might be integrated into the definition of social lifestyle. In fact social behaviors are partly defined by these contextual behaviors. For example, two people may meet while adjacently performing maintenance activities, or the social behavior may be dominated by a leisure activity like cards or tennis. Installation of a lock or placing barriers to screen barbeque tables while the family is eating reflect social preferences. The care of children often alternates with adult-adult social relations. In this perspective, social behavior is rarely "pure", i.e., uniquely involving the motive of being sociable and exclusively involving personal communications. In practice other kinds of behaviors go on coincidentally or alternatingly, like eating together, washing clothes together, caring for children together, playing games together, etc. Since these behaviors all imply physical design requirements it may be more valuable to define social behavior uniquely in terms of their contextual variables. For example, Becker found that the number of friends one has in multiple dwelling housing correlates significantly with residents' sense of security (4).

METHODOLOGY OF THE RESEARCH

Our methodology has been set forth in a previous publication (2). We use a variety of methods including: 115-question in-depth interview with housewives and couples; furniture maps of the unit; photography; tours around the environment and observations and fieldnotes recorded from the tour experience.

Within the interview 20 questions addressed the problem of social lifestyle. The questions were open-ended and called for short answers which were recorded directly on the schedule. A content analysis was performed on 8 questions which admitted of quantitative treatment of overall social potency for each sample project. This is reported below. Other questions were analyzed qualitatively and pooled with data from the other methods, particularly fieldnotes, in arriving at other dimensions of housing social lifestyle.

Our subjects, as required by government mandate, were limited to families (parent age range 25-47) with 2-3 young children (see Table 1 for age distribution) inhabiting 3-bedroom houses. The houses are low-density low-rise row and stacked arrangements consisting of units with and without basements, grade and stacked maisonettes and flats. Contact with subjects was organized through management in each case and the sample represents 80-90% of the total number of residents possessing potential sample requirements (size of family/size unit). The form of housing to be studied was also specified by government. Rents ranged from $140-175 in a uniform income population best described as "Moderate": $6,000-$12,000 a year.

Table 1. NUMBER OF FAMILIES AND CHILDREN BY PROJECT

Project	Language	Number of families	Number of Children	Children's Average Age (Years)
Mtl.-2	francophone	12	31	8.7
Sydney	anglophone	12	27	3.9
Ottawa	anglophone	12	30	7.3
Québec	francophone	12	28	5.1
	TOTAL	48	116	6.25 years

Each of the four projects is located in geographically illdefined areas, probably best described as an urban/suburban fringe or outer city. Within immediate view from the projects are open, vacant lands as well as other housing developments. Each project is also built near an expressway which runs into the center of town and close to a major shopping mall complex.

A total of forty-eight families were studied including twelve from each of the four projects (see Table 1) located in Montreal, Quebec City, Ottawa and Sydney, Nova Scotia.

Table 2. GENERAL SITE DENSITY AND HOUSING CHARACTERISTICS ALL PROJECTS

	M	O	Q	S
a. Total site area (sq. ft.)	544,025	313,500	204,630	384,942
b. Total site area (acres)	12.40	7.15	4.68	8.80
c. Construction at grade (sq. ft.)	121,862	73,375	50,339	76,198
d. Construction at grade (c/a %)	22.4%	25.0%	24.6%	19.8%
e. Total constructed area (sq. ft.)	302,290	123,946	199,120	152,800
f. Tot. flr. area/site area (e/a)	0.56	0.39	0.97	0.39
g. Types of units: 1-bedroom	66	---	30	---
2-bedrooms	43	---	54	---
3-bedrooms	163	118	62	50
4-bedrooms	40	---	4	54
h. Number of units/project	312	118	150	104
i. Number of families/acre	25.1	16.5	32.0	11.8

M - "Habitations Malicorne," Montreal
O - "Beacon Hill South," Ottawa
Q - "Ilôts Charlesbourg," Quebec City
S - "Rockliff Park," Sydney

SAMPLE PROJECT DESCRIPTIONS

Montreal (M)

The project "Habitations Malicorne," is located on the largest site (12.40 acres) of this study and was average with respect to density, containing 25.1 family units per acre. It is the largest project with regard to total number of houses having 312 units, and it has the greatest mix of house types varying from 1-4 bedroom flats and maisonnettes without basements and private outdoor space to walk-up 3-storey arrangements. The project is the richest with respect to community facilities indoors and on the site. Montreal has the following community equipment: an adult pool; four shops including a grocery, variety store, cleaner, and ice-cream store; two playgrounds; children's wading pool/sandbox play area; unplanned office space; benches; two municipally financed tennis courts which are across the street. The buildings contained laundry rooms. "Habitations Malicorne" is the newest project built in our sample and was only six months old when investigated. The residential sample consists of nine French-Canadian families and three English-Canadian families.

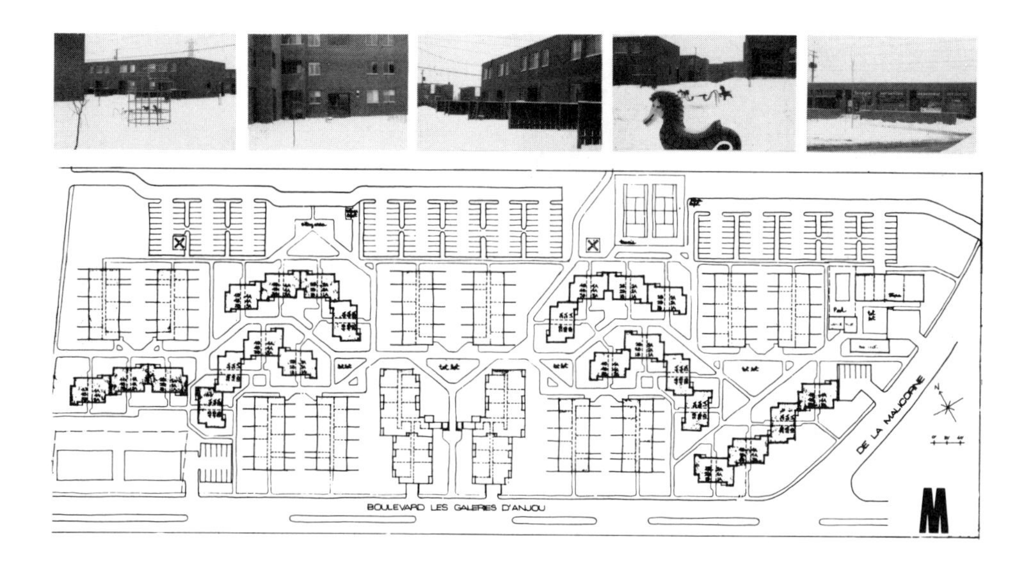

Quebec (Q)

The project "Ilôts Charlesbourg" is located on the smallest site studied (4.68 acres) and has the highest density of 32.0 family units per acre, in a total of 150 units. The project contains a relatively equal mix of 1-3 bedroom and a few 4 bedroom units in 2-storey row housing without basements and has fenced outdoor space for each unit. The project was relatively rich in its provision of community equipment. The project "Ilôts Charlesbourg" contained the following community equipment on the site: a hairdressing salon; general store; promenade; adult pool; badminton and volleyball courts; bowling green; and a horseshoe pit. "Ilôts Charlesbourg" is 4 years old. The residential sample consists of twelve French-Canadian families.

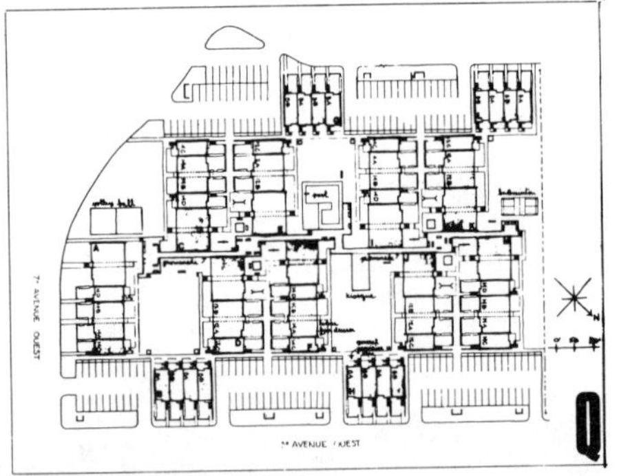

Sydney (S)

The project at Sydney, Nova Scotia - "Rockliff Park" - was built on the second largest site (8.8 acres) and contains the lowest density of any project being 11.8 family units per acre, and the lowest number of housing units, 104. It is a relatively homogeneous project consisting equally of 3-bedroom and 4-bedroom units in 2-storey row housing with basements but no fenced private outdoor space. The project had only one community amenity in the way of leisure equipment, a pool, and this was built by the tenants themselves two years after the project had been built. "Rockliff Park" is the second oldest project (5 years). The residential sample consists of twelve English-Canadian families.

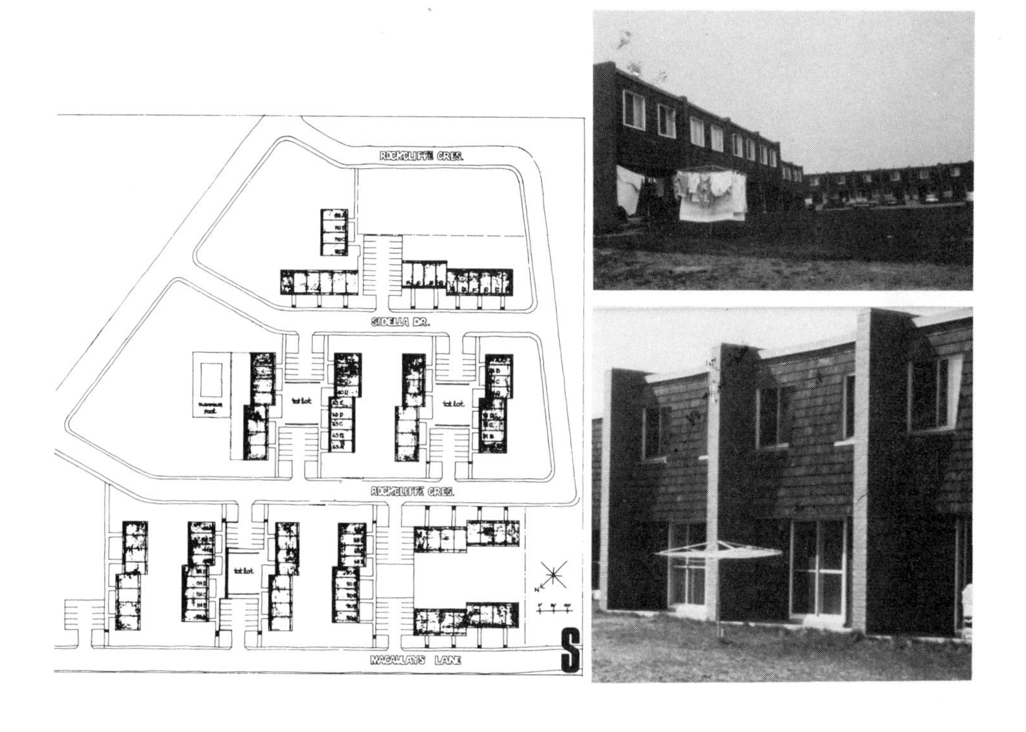

Ottawa (O)

The project "Beacon Hill South" was built on a relatively small site (7.15 acres) and contains the second lowest density, being 16.5 family units per acre and the second smallest number of housing units, 118. It is the most homogeneous project consisting uniquely of 2-storey 3-bedroom row housing with basements and private fenced outdoor space for each unit. The project was the weakest studied with respect to community equipment, containing only an adjacent playing field next to the project. No equipment per se was provided. "Beacon Hill South" is the oldest project of the group (7 years). The residential sample consists of twelve English-Canadian families.

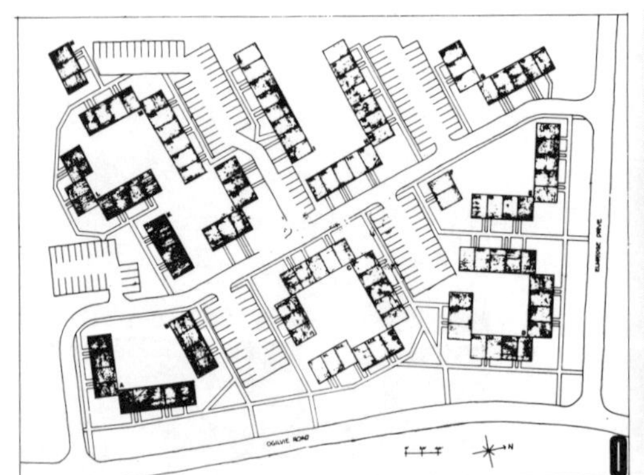

FINDINGS

Overall Potency of Social Life For Each Project

These overall findings are based on a combined analysis of responses to several questions for each project. A matrix structured nine questions on social life and produced scores which showed how extensive and in what form social life took place. The indicators and value ranges are as follows:

1. Number of shared activities, (0 @ 10)
2. Number of friends, (0 @ 9)
3. Number of *categories* of favors exchanged, (0 - 7)
4. Evaluation of community spirit, (-1 @ +1)
5. Definition of a good neighbor, (-1 @ +1)

6. Rate of entertaining, (0 - 8 per month)
7. Change in entertainment since moving to the project, (-1 @ +1)
8. Degree to which friends who were neighbors at previously lived in projects were retained as friends, (-1 @ +1).

Each of the indicators 1, 2, 3 and 6 received points based on frequency of mention in the interview, while variables 4, 5, 7 and 8 are ranked qualitatively.

The indicators were scaled and summed to yield total social life scores. The points assigned to each *activity* are derived empirically and we did not try to weigh the different factors, although recognizing that the factors are probably not equal in importance.

The scores are biased toward higher weights for indicators with larger ranges (1, 2, 3 and 6). Attitude scoring, such as for "definition of a good neighbor," had a narrow and low scoring range. We did not feel we could scale such findings on a larger range than utilized. The matrices are too extensive to reproduce in this paper.

Using such indices it was found that the four projects ranked as follows:

Table 3: OVERALL POTENCY OF SOCIAL LIFE BY PROJECT

	Number of Points	% of Total
Ottawa	231.5	38.4
Sydney	205	34.1
Montreal	102	16.9
Quebec	63.5	10.6
	602.0 points	100.0%

Ottawa and Sydney have more than twice as much social potency as the other two projects and the rate in Quebec is the lowest of all, contributing to only 10.6% of the total social life indices.

The score for Montreal, however, may be questionable in that the project had been occupied for less than a year and eleven of twelve residents sampled had been there less than 6 months. Another possible factor is that these residents are the only ones to live on upper floors. In addition, this sample consisted of somewhat older families with children, 2-3 years older than the average. Families with older children appear to have less occasion to run into each other, whereas mothers of pre-schoolers have a high rate of interaction.

The Quebec study group contained residents who had been in residence 2-3 years, the same as at Sydney, so their scores cannot be qualified in terms of length of residence. Other indicators which will be analyzed below will tend to confirm that the degree and kind of social life desired in Quebec was very different than in the other projects.

The Ottawa project, which has the highest scores, is also the oldest project, having been built in 1967. Residents in this project had lived there about a year longer than had any others and this may have influenced their scores, since they had more time to accumulate friends and engage in social activities.

The following conclusions are derived from the data concerning the overall potency of social life.

Entertaining

Residents were asked the number of times per month that they entertained visitors in the evening. The projects ranked as follows:

Ottawa	4.9 visits per month
Montreal	3.5
Quebec	2.5
Sydney	2.4

Among all respondents, Ottawa residents had experienced a sharp rise in entertaining since moving to the project. They felt that they had space to entertain more successfully than in previous residences. Sydney residents entertained less than they engaged in social activity in general, and it appears that the reason was largely economic; they could not afford to entertain much. Note that the projects in which entertaining is most frequent are both in larger cities.

Community Spirit

In spite of their high scores for social life in general, both Ottawa and Sydney residents felt that community spirit was relatively low in those projects. Montreal and Quebec residents evaluated their community spirit in a more positive light.

This appears to stem from two possibly very important and apparently related factors. One reason is that Ottawa and Sydney had the fewest community leisure facilities while Quebec and Montreal had rich community facilities. Moreover in Sydney, the one facility, a pool which had been developed by a tenants' association, was the object of much management/resident and resident/resident dissension because of financial and overcrowding problems. Another factor concerns the possible inverse relationship that exists between interpersonal neighboring with proximate neighbors and participation in community activities in the context of community equipment. Perhaps interpersonal close neighboring "uses up" the desire for other kinds of social contacts. Perhaps there is a disposition to neighbor privately in someone's home more than in public. Thus is Ottawa and Sydney, most social life was absorbed in functional exchanges and light sociability with immediate neighbors while in Quebec and Montreal most contacts took place in the context of community facilities.

The facilities which seem most efficacious in this regard are the pool, commercial stores and the promenade. Community rooms which lack definition and equipment tend to be very underused.

Favor and Service Exchanges (Help)

Ottawa and Sydney residents had engaged in a total of 4.4 categories of favors given and/or received since living in the housing project. Favors included borrowing and/or lending of almost every kind of household food, tool or product, as well as babysitting, rides to work, and shopping. Quebec and Montreal had engaged in only 1.8 favor categories.

However, it appeared from other evidence that Montreal mentioned "Help" as part of a definition of good neighboring. Quebec residents, although mentioning help, had a large percentage who said that the best neighbor is someone who minds his own business and who does not impose on others. Although we received information on the categories of favors given and received, we have no firm information on how frequently such activity takes place.

Shared Activities

Shared activities meant mutual visiting, going shopping together and caring for children together. Concerning this dimension of social life, the projects ranked as follows:

Sydney	4.2 kinds of activities
Ottawa	3.9
Montreal	2.0
Quebec	1.5

Again Ottawa and Sydney are consistently higher than Montreal or Quebec. From qualitative as well as quantitative evidence it appeared that in Sydney residents engaged in significantly more dropping in on each other and performed more joint activities than in Ottawa which was more oriented toward helping activities. Quebec residents, characteristically, score lowest on shared activities with other residents.

Number of Friends Per Resident

This concerns the number of other residents in the project called "friends." Sydney had 3.3 friends per resident, Ottawa had 3.0, while Quebec and Montreal had only 1.5 friends per resident. Thus the friendship rate in Sydney and Ottawa is somewhat more than twice as much in Quebec and Montreal.

How Residents Met Their Friends

Ottawa: (all residents had friends - range 2-5 friends).

In the Ottawa project, residents met in primarily two ways: 1) while sitting out in front or behind their houses or somewhere unspecified on the project grounds; 2) through their children: in this case the children meet first, then children visit each other until mothers sort of run into each other engaged in picking up children from the incipient neighbor's house or in the context of watching children play. Another way in which residents meet, as mentioned by two families, consists in a spontaneous juxtaposition of maintenance activities in which both residents happen to be mowing the lawn at the same time or shovelling snow and just get to talking. There were other possible means of meeting among the families studied: through introduction by other residents; through distant family ties; through husbands holding jobs in the same organization. One resident was an "Avon Lady," and her job brought her into contact with a great many residents. Thus, there are both physical and social encouragements to residents meeting initially. Residents may simply be in the same space at the same time, doing a similar activity (e.g., sitting out, mowing the lawn, getting out of their car) and may just bump into each other and start interacting. Or residents may be introduced through the liason of children, who with their facility for making friends act as socializers for the parents. More formal introductions also occur through family ties, mutual friends and related jobs.

Sydney: (all residents had friends - range 1-9 friends).

In the Sydney project, the most frequent means of meeting was in the immediate outdoor space around the house. Meeting while sitting out in the back was common or when watering the grass. We may note also that every two adjacent residents shared a common water tap and this might have promoted a mild bond of contact between them. Residents here meet through children to a lesser extent than in Ottawa; children were a greater source of antagonism in Sydney (littering, vandalism) than in Ottawa. As in Ottawa some residents met through introductions by mutual friends, or by husbands holding the same job. Interestingly, there was another mention that a commercial activity, in this case a Tupperware Party, led to the resident meeting their friends for the first time. Sydney residents primarily meet informally and spontaneously in the outdoor space immediately adjacent to their units. Three women mentioned that they met "by walking over and introducing myself." We might suspect that there may be some cultural differences showing up here, to wit that Maritimers might find it easier to make friends with neighbors, might be less formal with them and more willing to take the initiative given the fact that they are already spatial neighbors.

Quebec: (seven of twelve residents had friends - range ½[2]-2 friends).

In analyzing social life in Quebec, we are using a smaller number of existing friendships as an information base. Only two families report meeting outside in the immediate vicinity of their house, one while sunbathing. Another two families say they met friends while strolling around the promenade that runs through the project. Other families had been introduced through a mutual friend, husbands' jobs or by the concierge.

A couple of families say they met through their children. Thus we continue to see a great variety in the means by which residents come to meet their friends. Although Quebec only has seven families reporting (five families reported no friends in the project) there is nevertheless a great variety in the way in which people meet.

Montreal: (seven of twelve residents had friends - range ½[2]-3 friends).

In this project, a relatively small number of families professed to have friends in the development. Two families met through children, two while sunbathing outside the house, two when both women

---

[2] Mild friends

were washing windows, one had a family tie, and two because both husbands came from the same town. As noted, one family knew an Avon Lady who met friends through her job. As in Quebec we find a very wide range of sources of meetings but there is clearly a lower degree of association than in Ottawa or Sydney. There were no mentions of meeting at community facilities, though these may have been used to sustain contact.

Definition of a "Good Neighbor"

Residents of all projects were asked to describe the qualities of a good neighbor. In each project a certain number of persons responded to this question with "someone who minds their own business" or words to that effect or "someone who does not impose and does not approach you." As a preliminary analysis we have examined this "mind their own business" response in each project. We believe this question elicited social values which govern actual or potential interactions among residents:

Table 4: "MIND THEIR OWN BUSINESS" RESPONSE: ALL PROJECTS

Montreal	12 out of	26	total responses	46.6%
Quebec	11 out of	25	total responses	44.4%
Sydney	5 out of	25	total responses	20.0%
Ottawa	3 out of	20	total responses	16.7%
Grand Total:	31	96	Aver.	32.2%

It is apparent that among Ottawa and Sydney residents there are much more positive attitudes toward neighboring.

In Ottawa, the definition of a good neighbor and good neighboring was largely (8 of 20) that of someone with whom you exchanged favors and borrowed and lent products. The range of the individual family responses are as follows:

"people with whom you can trade favors" - three families
"exchanges of favors" - two families
"can borrow from" - one family
"someone who does not borrow too much" - one family
"someone who returns borrowed things" - one family

A lesser number defined good neighbors as people who were good company and whom you could talk to. One family mentioned someone who keeps a good living standard, and two mentioned "co-operative" as a good quality.

In Sydney, the leading quality of a good neighbor was "friendliness," mentioned 11 of 25 times; some qualified this by saying "not too friendly." Other residents defined good neighbors as informal, co-operative, "non-complaining," and there was one mention of "someone who is quiet."

In Quebec, 44% defined the good neighbors as: "someone who minds their own business." Of the other 56%, the majority called the good neighbor the "one who can help you and whom you can help."

The definition of help was generally addressed to emergency situations by two of the families: "someone you can count on," "someone who would let you use their telephone in an emergency." Some families used positive definitions, such as people with whom you could talk and be sociable.

In Montreal, of the 54% who did not answer "minds own business" to this question, the majority (7/26 mentions) used words which we may paraphrase as "sociable," "amiable," "welcoming," "approachable." Two families mentioned tolerance and understanding of children. Three families mentioned help-related definitions of good neighbors and two specifically mentioned "discrete" and "does not gossip."

In summary, we may characterize each project as follows with respect to the kinds of qualities which dominated descriptions of good neighbors:

Ottawa:	exchange of favors and borrowing;
Sydney:	friendships and readiness to help and borrow;
Quebec:	minds their own business and readiness to help in emergency;
Montreal:	minds their own business and sociable.

There were five mentions of tolerance for children in all projects but Sydney. Helping and borrowing definitions comprised 27 of 96 responses, making this the second leading definition of a good neighbor after "minds own business." Each project seems to have a distinct style of approach to neighboring: on the one hand the Quebec and Montreal projects had relatively negative and functional (help) definitions; while Ottawa and Sydney had functional, sociable and friendliness orientations.

Social Lifestyle and Family Type
(Stage of the Family Life Cycle)

In this analysis families are categorized[3] according to stage of the family life cycle and then young types and older types compared. In this case we compare social lifestyle among Type I (N = 14) who have at least two children under the age of 5 and one is an infant of less than 18 months; Type II having two children between 5 and 9 years (N = 26); and Type III (N = 8) who have older children (10 to 18 years). Neighboring scores were computed for the forty-eight families. Scores represent the average number of social potency points per family (see Table 2).

Table 5: SOCIAL LIFESTYLE POTENCY BY FAMILY TYPE

Family Type	Number of Families	Family Type Average Potency Score
I	14	13.2
II	26	14.4
III	8	7.2

In view of the small number of Type III (the group with older children) we may come to qualitative but not statistical conclusions concerning the relationship between stage of the family life cycle and intensity/complexity of neighboring. However, it appears that Type III families do have less social or neighboring activity than the younger Types I and II who have children from infants to 9 years old. The trend is in the same direction as an earlier finding in the pilot project where young families ranked much higher in potency of social life than older families with teenage children (3).

In other data (other interview questions) we see further proof that younger families have different *attitudes and perceptions* of the project as well as actually engaging in more neighboring activity. In their view, there should be more social contact among neighbors and they are more likely to believe that residents know each other quite well. Families with older children tend to feel that neighboring should be limited and that their neighbors knowledge of each other is and should be limited in number and intensity.

This is leading us to the formulation of criteria for the mix of families in any one project region (be it row, or court, or building grouping in scale). From the evidence thus far collected, we would lean toward a plan which allowed young families to live in

[3] Family type categorizations are discussed in detail in "The Evaluation of Family Satisfaction with the Design of Stacked Masionnette" (2).

one area of the project and older families in another. We hypothesize this on the grounds that both quantity and quality of social life would be potentially more harmonious among families of a particular stage: that older families desiring less social life would find more outgoing young families somewhat more imposing. We are not suggesting, however, projects devoted uniquely to any one family type.

Neighboring and Proximity

In almost all cases residents neighbored primarily with others who lived proximately. In row housing this amounts to saying that the bulk of neighbors come from within one's row and to a lesser extent from visually proximate neighbors who lived across the way. Friends, however, are not necessarily proximate neighbors, since they may result from social introductions, from children who play and make friends from all over the project, or considerations other than physical and/or visual proximity. Neighboring within a row may extend also to houses which are around a corner.

Examples: In Sydney, there was a row of houses (Sidella Drive) which was separated by a public street from the rest of the project. This row of 10 houses produced a great intensity of neighboring among nine of the families in two distinct neighboring groups. The 10th family had just recently arrived.

In Ottawa, there were 2 rows of housing containing a dozen houses fronting on each other across a minimum standard 35-foot separation. These houses had many young children as well. The result was a tremendous amount of social activity among neighbors. One family reported she babysat for other families without leaving her house. It is hard to evaluate the costs as well as the benefits of a highly interacting social matrix such as this. A family which valued its privacy would be quite unhappy in that situation. Some (young) families are accustomed to the noise of young children whereas older families or families with more privacy orientation are not as tolerant and this can lead to friction.

In Quebec, the investigator concluded that most neighboring was initiated by mothers looking after children, by fathers and sons mixing in leisure sports exchanges with other fathers and sons, and by some women around the pool. Here proximity was not a strong factor in promoting neighboring; on the contrary, one kept one's distance from those closest.

In Montreal, the investigator concluded that most neighboring was related to propinquity.

In Montreal-1, (a project not included in data presented heretofore) neighboring occurred among residents whose patios backed into an open court; but not across the hall with neighbors who were proximate but whose patios faced front. The bulk of neighboring was initiated through informal leisure time contact in the immediate vicinity of the outdoor patio (3).

These findings need to be followed up with additional research involving a locational sociometric approach such as that of Carey and Mapes (5).

DESIGN RECOMMENDATIONS

Private and Public Outdoor Space Adjacent to House

On this subject we have some ambivalence, but in general support the idea that every family is entitled to a private outdoor space, that if the family has a private outdoor space it will engage in outdoor activities more, activities which involve family leisure and relaxation. For families with strong privacy orientation, a private outdoor space is one factor which plays a very important role in overall satisfaction with housing.

Nevertheless, we suspect that, although there is some evidence to the contrary, providing everyone with a private outdoor space would probably reduce neighboring, but to what extent in quantity or quality we have, of course, no idea. In Sydney people neighbored because they sat out in adjacent spaces with no barriers in the way of walking across an open space to relate to a neighbor. In Ottawa, people neighbored because they sat out in front of their houses, in spite of the fact that they had fenced patios. They preferred to sit out where they could see other people.

We recommend that houses be provided with a relatively public space on one side of the house and a relatively private space on the other. We know of two projects (Sydney and Montreal-1) where an arrangement that provided for patios facing each other (in rear) were relatively unsuccessful. In Montreal-1 the patios faced inner courts, and this may have provided too much exposure (3). In Sydney, one row of unfenced patios was located opposite another row of patios, but children and dogs interfered with patio activities.

Patio activities were regarded as more private than sitting out in front, near the front door. It appears from the Ottawa project that families will meet and neighbor out in front of their house, but prefer seclusion to the rear of their house. This very much resembles the human posture when we make contact with others, face-to-face, and when we want privacy we have but to turn around. The front is also partially regulated through the facade and the garden; through these elements one is judged by neighbors with respect to standards of decor and maintenance.

With respect to fencing it would appear that residents should be given the option of fencing in a portion of their adjacent open space to the rear of their house. Room should be left to allow a path parallel to the house, i.e., a path by which one can still have access to a neighbor somewhere on the other side of fenced rear spaces.

Community Equipment: Social Contact Initiation

Housing projects should provide a wide range of community activity spaces and services to provide residents with a variety of opportunities to come together to pursue common purposes. While less defined spaces such as open spaces, benches, paths are necessary to support or provide access to special function areas, community equipment is directly conducive to social contact initiation. Such areas should be easily accessible, functionally well equipped and comfortable - providing support elements (i.e., seating, lighted areas) for those who prefer to watch and talk rather than participate.

Family Type Mixing

In view of the number of families advocating some spatial division between families with young children and the elderly, it seems reasonable to recommend that housing projects offering both family (multibedroom) and non-family (few-bedrooms) housing types should cluster each type separately - without preventing elderly or childless couples to rent family type housing if they so desire with the foreknowledge that they will be located in a family cluster.

The situation with respect to mixing young families and older families with teenage children is less clear than for families with children and singles or elderly couples. We think experimentation and further research is necessary before definitive recommendations can be made. However, we believe as a general principle that in a new project where the procedure is feasible, families of similar types should be placed as proximately as possible. Subsequently, new families coming into the project should be made aware of this arrangement and informed of the kind of social milieu into which they are being placed. We are aware that probably only an experimental project can really carry this program out fully.

Architect or Designer Requirements

Architects should be aware of their own dispositions for privacy or social interaction since they may unconsciously impose their standards on others. Individual designers are bound to have unique orientations based on culture, personality and environmental experience. Man-environment sensitivity training for various groups should be mandatory in the training of those who give us our environments.

SITE SOCIAL REQUIREMENTS SYNTHESIS

From the data and from our prior formulations of the problem it is very apparent that social relations in the context of housing comprise a system of great complexity. It is not desirable to prescribe definitive design conclusions on the basis of this data. Information should continue to be generated by looking at different forms of social interaction taking place in different locations at different rates of frequency.

We find among diverse population groups that highly interactive neighbors tend to live in relative propinquity to one another. Whether one neighbors fore or aft is determined by which space is "sat out in" more. In the low densities studied, the neighboring pattern appears to be favorably increased if two families are living in the same building, even around a corner. However, social relations, notably friendships, may be found regardless of location, especially when friends first meet through a social introduction or through children who range more widely.

It is also important to analyse the way people first meet and the way in which relationships are sustained. The most powerful early factors among two potential neighbors appear to be house proximities, children who play with one another, or mutual friends. We really have no knowledge whether these early factors exert some influence over subsequent routine rates of neighboring. Both community equipment and the potential for sitting out within mutual view appear to be important sustaining factors for social relations.

The social system is apparently governed by factors of *culture*, in the sense that crucial French-English differences have emerged; by factors of *personality*, in the sense of the "mind your own business" or privacy syndrome; and by *demographic* factors, in the sense of the family type definitions using size of the family and stage of the life cycle. It is clear that one needs to look at both intra-individual relations as well as inter-individual relations. Because the populations studied were carefully matched no socio-economic class differences emerged here.

One of our important conclusions is that different cultures may well have different design requirements concerning the promotion of social relations. The sample of French families in Quebec and the Montreal project differ both quantitatively and qualitatively from the sample of English families in Sydney and Ottawa in the scope and desire for contact with neighbors. This of course would need confirmation through additional studies and only refers to certain kinds of families such as those studied. Should such a trend be confirmed in additional research it would imply that projects built in Quebec and French Montreal might need fewer spaces for potential interaction with near neighbors, perhaps more community facilities as nodes of interaction, and greater care with respect to privacy around the unit itself (fencing, windows, visual barriers, soundproofing, individual entry, etc.).

Another important finding is that the English cultural pattern of social interaction is defined to a great extent in functional (use) terms, with the majority of the activity taking place in helping - borrowing and lending of goods and services. In such a case one does not want to place too many privacy barriers among a group of proximate neighbors. The English residents, although their projects had little community equipment, did not complain about this except to express a desire for more facilities for children. Thus we may tentatively say that the English neighbor in more proximate building groups while the French neighbor more in community facilities away from their houses. The social drive is eqully strong but gets expressed in ways unique to each culture. There also is the possibility that community facilities in the French projects may have attracted social contacts in those environment and limited proximate neighbor contacts. This idea is based on the intuitive notion that there is a ceiling or range of social energy available to each resident and if it is absorbed in the community spectrum, there may be insufficient energy or potential for contacts in the proximate zone.

Because social needs in different cultures may vary we would think it desirable that designers be members of the client's cultures where possible.

We are concluding that continuing attention should be devoted to the question of family type matching as this is related to the construction of variously sized units and their distribution through the project. The information indicates that resident families probably prefer to live among similar types, e.g. (size and stage of life cycle). But the tendency to homogeneity is not as narrowly expressed as we had supposed. Young families with young children can coexist with older families with teenagers. Babysitting is an example of an activity in which the families are complementary, and there are other complementary relationships possible (older family counseling younger, etc.). However, on both sides there seems to be a definite preference not to be located in the same area. It is significant that Ottawa, which had the highest interaction rate in social relations, was also the most homogeneous with respect to house type (118 units: all 3-bedrooms). Moreover, we can point to about five or more older families in our sample who were the least satisfied with their housing, for a variety of reasons, but definitely including social factors.

Another important factor in resident social life appeared to be personality orientation. About 1/3 of all residents queried defined a good neighbor as "someone who minds their own business." A certain number of the responses can only be attributed to distinct antisocial tendencies in the housing context. To a certain extent this is mitigated by their social relations with outside friends or family but this is not always the case. The designer must expect that a certain not insubstantial proportion of families are antisocial with respect to fellow residents.

A conclusion related to the above is that there is a highly important role for child interactions in stimulating adult meetings which lead to friendships. We found that children were often intermediaries in adult meeting situations in housing. We might hypothesize that families with children of similar ages would have more in common than families with children of dissimilar ages. The more similar the ages, the more likely children will be to interact among themselves. It is likely also that families with children of different ages have different tolerances for the problems associated with each age group. This may be especially true with respect to noise levels. A family with teenagers might understand the habits of teenagers more sympathetically. All this supports the idea that families in similar life stages tend to interact with and might prefer to reside near families of their own type.

REFERENCES

1. Altman, E. Privacy: A conceptual analysis. In S. T. Margulis (ed.) *Privacy*, Part 6, Proceedings of the Fifth Conference of the Environmental Design Research Association, Milwaukee, 1974.

2. Beck, R. J., Rowan, R. and Teasdale, P. The evaluation of family satisfaction with the design of the stacked maisonette. In C. Lozar (ed.) *Methods and Measures*, Part 5, Proceedings of the Fifth Conference of the Environmental Design Research Association, Milwaukee, 1974.

3. Beck, R., Rowan, R. and Teasdale, P. *Phase 1 Report*. Ottawa: Central Mortgage and Housing Corporation, 1974.

4. Becker, F. D. The effect of physical and social factors on residents' sense of security in multi-family housing developments. *Journal of Architectural Research*, 1975, *4*:18-24.

5. Carey, L. and Mapes, R. *The Sociology of Planning*. London: B. T. Batsford, 1972.

6. Margulis, S. T. Privacy as a behavioral phenomenon: coming of age. In S. T. Margulis (ed.) *Privacy*, Part 6, Proceedings of the Fifth Conference of the Environmental Design Research Association, Milwaukee, 1974.

7. Perin, C. *With Man in Mind*. Cambridge: M.I.T. Press, 1972.

8. Wolfe, M. and Laufer, R. The concept of privacy in childhood and adolescence. In S. T. Margulis (ed.) *Privacy*, Part 6, Proceedings of the Fifth Conference of the Environmental Design Research Association, Milwaukee, 1974.

DETERMINANTS OF HOUSING PREFERENCE IN A SMALL TOWN

H. Jaanus and B. Nieuwenhuijse

University of Amsterdam, Subfaculty Psychology
Weesperplein 8
Amsterdam, The Netherlands

INTRODUCTION

In the present study an attempt was made to determine preferences for certain residential areas in a small Dutch town and to compare these preferences with the subject's appraisals of various physical and social aspects of these areas. Dutch residential areas are suburban districts which, especially in modernized towns, form real architectural units. While they differ among themselves, each of them is homogeneous as far as spatial design, type of housing, and provision for green zones are concerned.

Many aspects of a suburban district as appraised by a subject might influence his preference to live there; to mention a few: his appraisal of the quality of the spatial design, the quality of housing, the privacy these houses guarantee, shopping facilities, recreational and social facilities, etc. It is difficult to assess beforehand which aspects will determine housing preference; is it those aspects which are associated with privacy, or social functioning, or is it those associated with green zoning? In studies such as these a broad spectrum of characteristics of residential areas should be assessed (see Table 1) in order to reduce the chance of overlooking the really effective determinants of housing preference. The more general appraisal dimensions may then be determined by, e.g., factor analysis. For example: the degree to which a district offers shopping facilities, promotes social contacts, etc., may quite possibly have such coherence that one can demonstrate the existence of a single underlying appraisal dimension (factor), which in this case might be called 'social functioning.' Should the large number of appraisals of different suburban aspects be reducible by

Table 1: ASPECT RATINGS AND SHORTENED INDICATIONS USED (Ratings are ranked on a 7-point bi-polar scale)

Number	Scale	Indication
1	How would you feel about living in this area? (poles: pleasant-unpleasant)	Housing preference (pleasant-unpleasant)
2	What do you think of the green zones (poles: imaginative-unimaginative)	Green zones (imaginative-unimaginative)
3	Do you think there are sufficient green zones in this area? (poles: sufficient-insufficient)	Sufficient green (sufficient-insufficient)
4	What do you think of the appearance of the green zones? (poles: abundant-sparse)	Appearance green (abundant-sparse)
5	What do you think of the spatial design of this area? (poles: spacious-cramped)	Spatial design (spacious-cramped)
6	What do you think of the position of the houses in relation to each other? (poles: imaginative-unimaginative)	Position of houses (imaginative-unimaginative)
7	Has the road construction added to variety in this area? (poles: greatly-little)	Road construction (greatly-little)
8	What do you think of the street plan in this area? (poles: clear-not clear)	Street plan (clear-not clear)
9	Would you expect much nuisance value from traffic on the streets? (poles: little-much)	Traffic nuisance (little-much)
10	How do you rate this area in terms of traffic safety? (poles: safe-unsafe)	Traffic safety (safe-unsafe)
11	Do you think this area is safe for children (traffic; water)? (poles: safe-unsafe)	Children's safety (safe-unsafe)

cont'd....

12	Do you think this area is a pleasant area for children to play in? (poles: pleasant-unpleasant)	Children's play (pleasant-unpleasant)
13	Do you think traffic in this area will cause much indoor noise annoyance? (poles: little-much	Indoor noise annoyance (little-much)
14	Do you think one's neighbors will form a problem in this area? (poles: negligible-considerable)	Problem neighbors (negligible-considerable)
15	Do you think one's neighbors will be a source of noise annoyance in this area? (poles: little-much)	Neighbor noise annoyance (little-much)
16	Do you think privacy would be invaded by neighbors and passers-by? (poles: little-much)	Privacy invasion (little much)
17	Do you think this area is built in such a way that one can lead one's own life? (poles: very much so-not at all)	Privacy (very much so-not at all)
18	Do you think this area provides enough private space around the house? (poles: much-little)	Private space around house (much-little)
19	What is your visual impression of the area? (poles: tasteful-ugly)	Visual impression (tasteful-ugly)
20	Is nature represented in the area? (poles: greatly-not at all)	Representation of nature (greatly-not at all)
21	What do you think of the liveliness of the street scene in this area? (poles: lively-dull)	Street liveliness (lively-dull)
22	Do you think this area provides enough shopping facilities for daily necessities? (poles: sufficient-insufficient)	Shopping-daily necessities (sufficient-insufficient)

cont'd...

23	Do you think this area provides enough shopping facilities for goods of lasting quality? (poles: sufficient-insufficient)	Shopping-goods of lasting quality (sufficient-insufficient)
24	What do you think of the design of this area? (poles: varied-monotonous)	Design area (varied-monotonous)
25	Do you think this area provides enough outdoor recreational facilities? (poles: sufficient-insufficient)	Outdoor recreational facilities (sufficient-insufficient)
26	Does the area strike you as being particularly colorful or colorless? (poles: colorful-colorless)	Impression of area (colorful-colorless)
27	Do you think this is a pleasant area? (poles: pleasant-unpleasant)	Pleasantness of area (pleasant-unpleasant)
28	Do you find the area isolated from social and cultural centers? (poles: not at all-considerable)	Social isolation (not at all-considerable)
29	Are there enough schools in the area? (poles: enough-not enough)	Provisions of schools (enough-not enough)
30	Do you think this area provides enough facilities for recreation (restaurants, pubs)? (poles: enough-not enough)	Recreation facilities (enough-not enough)
31	Is there enough parking space in this area? (poles: enough-not enough)	Parking space (enough-not enough)
32	What do you think of the distance between the blocks of houses? (poles: pleasant-unpleasant)	Distance between blocks (pleasant-unpleasant)
33	Most residential areas in this town differ greatly in age. How do you evaluate this area in this respect? (poles: pleasant-unpleasant)	Age of area (pleasant-unpleasant)
34	Are the houses luxuriously or simply constructed? (poles: luxurious-simple)	Construction of houses (luxurious-simple)

cont'd...

35	What do you think of the quality of the houses? (poles: good-bad)	Quality of houses (good-bad)
36	Does the construction of the houses in this area strike you as expensive or cheap? (poles: expensive-cheap)	Value of houses (expensive-cheap)
37	How well do you know this area? (poles: very well-not at all)	Familiarity with area (very well-not at all)

factor analysis to a small number of appraisal dimensions, then one could predict housing preference from this, providing a more comprehensive result.

In anticipation of a detailed discussion of the design and the results of the present study, we may say at this point that such appraisal dimensions were indeed found in our study. Housing preference was predicted from the three most important appraisal dimensions, namely, 'comsit','social functioning', and 'green zoning provisions'. 'Comsit' (a combination of 'comfort' and 'situation') refers to all sorts of aspects of housing accomodation and spatial situation of the houses. Both aspects are closely linked to all forms of privacy, visual and auditory as well as territorial privacy. Futhermore, it can be demonstrated that this dimension is by far the most important predictor of housing preference in contrast to the other two (see figure 2). Since generalizations on the basis of results from studies such as these should not be made indiscriminately, it should be noted that the town in our study, Uithoorn (situated 20 kilometers from Amsterdam, with approximately 12,000 inhabitants) has a predominantly middle class population whose way of life is typically Dutch and thus, northern-european. The results of this study may not therefore be generalized to e.g., metropolitan areas or to towns with a predominantly working-class population. Furthermore, one could very well expect that diametrically opposed results would be obtained from a study concerned with a population whose way of life is typically mediterranean. The social functioning of districts (a factor which includes such aspects as sociability and liveliness of street life as well as shopping facilities, recreational possibilities, etc.) might probably be a much more important determinant of housing preference in a mediterranean society than e.g. privacy. The results of this study do not imply that 'social functioning' is a factor of secondary importance in general; they imply solely that 'social functioning' is not an important predictor of housing preference. Obviously, this finding does not belie its importance as a factor in a countless number of other aspects of life in a suburban district. It goes without saying that here too generalization from these results to towns with another way of life should be made with great caution.

The use of residential areas as appraisal objects in the design of this study is, in our opinion, new. Methodologically speaking, the present study has much in common with other studies (e.g., 1, 3, 4, 6). It differs from these in the nature of the socio-planological units which were appraised, namely, housing units for one family in suburban districts versus buildings, apartments in metropolitan districts. In Uithoorn the opinions of the inhabitants with regard to the various residential areas in the town were obtained on subjects such as quality of housing, privacy, variety in design, shopping facilities, provision for green zones, road safety, etc. After a factor analysis of all these impressions of the subjects was carried out the subject's willingness to live in a given residential area was related to the factors.

At the time of the study, Uithoorn consisted of eight different residential areas (see fig. 1), clearly separated by water, railway embankment or road. The oldest of these areas, the only pre-war one, is situated on the river Amstel. It has a shopping center and in general, all the characteristics of a small town. The other areas stretch out in a ribbon development. They are numbered in the order of construction in Figure 1 and Table 2. In the most recently-built area, still under construction, No. 8, several ten-storey flat buildings have already been completed. In the eastern sector of Uithoorn an industrial area stretches along the Amstel. Otherwise, structures are predominantly low-rise buildings.

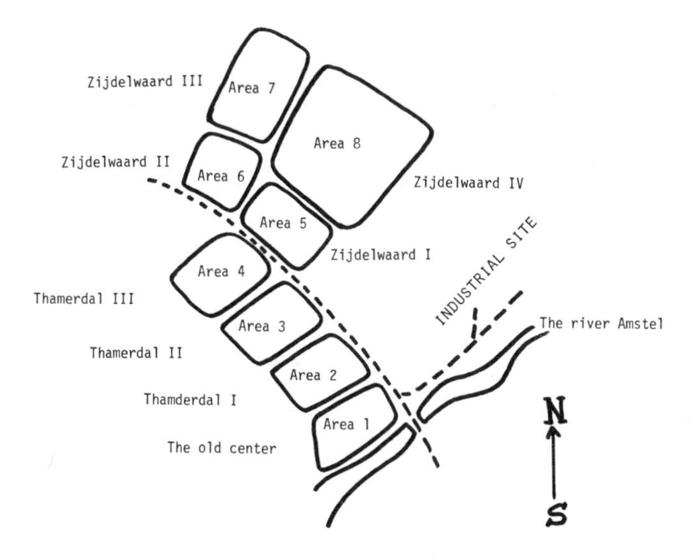

Fig. 1. Plan of Uithoorn.

AREA 1: THE OLD CENTER

AREA 2: THAMERDEL I

AREA 3: THAMERDEL II

AREA 4: THAMERDEL III

AREA 5: ZIJDELWAARD I

AREA 6: ZIJDELWAARD II

AREA 7: ZIJDELWAARD III

AREA 8: ZIJDELWAARD IV

PROCEDURE

Subjects

A random sample of eighty residents, spread evenly over the eight areas under study, were asked for their evaluation. The only restrictions applied were an equal division into age groups 18-35 and 36-65, and an equal number of men and women per area.

Material

Eighty participants in the study were shown eight color photos of Uithoorn residential areas, including their own, before being asked to give an appraisal. Each subject was also shown the position of each area on a colored map of Uithoorn and he was then asked to give his appraisal with regard to a total of 37 aspects, identical for each area (Table 1). The eighty participants thus appraised each of the eight areas for 37 aspects. The first of the aspect ratings (No. 1, Table 1), called the 'housing preference scale', concerned the subject's willingness to live in a given area. The remaining 36 aspect ratings, thought to have some correlation with housing preference, were partly selected on the basis of a study by Peterson (4) in which photos of buildings were used. We also followed a factor analytic study by Franke & Bortz (2) which placed its main emphasis on the search for dimensions in the connotative appraisal of residential areas.

The appraisal scores represented three dimensions: the fact that the scores with regard to housing comfort, situation of the houses in relation to each other, and the amount of privacy appeared to be of equal importance, resulted in the formulation of a dimension called "Comfort and Situation" (Comsit). The other two dimensions distinguished were "Social Functions and "Green Zones," pertaining respectively to opportunities for social contacts (recreation, shopping, and so on) and to the green zone provisions in the residential areas.

RESULTS

The total average scores per area for the eighty subjects were calculated. The average housing preference scores are shown in Column 4, Table 2. It appears that the old town center was not highly rated as a residential area. Two of the older areas (Thamerdal I and II) also received fairly unfavorable scores, as was the case with the high-rise area. However, the old area obtained a favorable score on two of the aspect ratings, No. 21, 'liveliness of street scene' and No. 23, 'opportunity of buying goods of lasting quality'.

Table 2. DIMENSION SCORES OF THE 8 RESIDENTIAL AREAS WITH SCORES ON THE HOUSING PREFERENCE ASPECT RATINGS (All scores are standard scores)

Number of area	Name of area	Dimension scores: Comsit	Social functions	Green zones	Housing preference score
1	Old center	- .78	+2.45	- .78	-1.28
2	Thamerdal I	-1.32	- .32	+ .58	- .91
3	Thamerdal II	-1.19	- .55	+ .61	- .99
4	Thamerdal III	+ .08	+ .08	+ .84	+ .33
5	Zijdelwaard I	+ .66	- .54	+ .46	+ .75
6	Zijdelwaard II	+1.38	+ .17	+ .57	+1.22
7	Zijdelwaard III	+1.21	- .19	+ .04	+1.44
8	Zijdelwaard IV	- .09	-1.10	-2.33	- .56

* * * * * * * * *

Table 3. THE CORRELATION MATRIX OF ASPECT RATINGS CLOSELY RELATED TO HOUSING PREFERENCE

Number	Shortened indication	Correlation coefficient
5	Spatial design (spacious-cramped)	+.95
6	Position of houses (imaginative-unimaginative)	+.93
7	Road construction (great variety-little variety)	+.94
17	Privacy (very much-not at all)	+.95
18	Private space around house (much-little)	+.97
19	Visual impression (tasteful-ugly)	+.92
26	Impression of area (colorful-colorless)	+.95
32	Distance between blocks (pleasant-unpleasant)	+.95
33	Age of area (pleasant-unpleasant)	+.94
35	Quality of houses (good-bad)	+.97

The Selection of Dimensions on the Basis of the Data

A factor analysis was applied to the scores obtained for the appraisal aspects to determine one or more dimensions which could be used directly without having to take all the separate aspect ratings into consideration.[1] This was based on a correlation matrix showing which of the 37 scales correlate with the 'housing preference scale'. Table 3 indicates those correlations with housing preference which have an absolute value higher than 0.90.

Results of the Factor Analysis

The factor analysis was applied to 36 aspect ratings, the housing preference aspect being left out of consideration. Three dimensions, D1, D2, and D3, were found. Table 4 shows the factor loadings of the 36 scales for the three dimensions. All loadings between +.30 and -.30 have been omitted. The results indicate that the three dimensions taken together adequately cover the whole range of the material; the reduction of the large number of aspect ratings to a few dimensions may be said to have been successful. Finally, it can be noted that the significance of the dimensions diminishes from left to right.

Dimension 1

Let us first concentrate on the aspects which have high loadings on Dimension 1 (D1, Table 4). Some of the aspect ratings are clearly concerned with the quality of the dwellings (aspects 33-36), whereas others pertain to aspects of privacy (aspects 14-18). All forms of privacy are presented in this first dimension, i.e. visual privacy (aspect 16), auditory privacy (aspect 15) and territorial privacy (aspect 18). Other aspects pertain to the structure of the area (aspects 5, 6, 7, 24, 32 and again 18).

The first dimension seems to present a typical 'town planning' aspect, i.e. spatial structure of situation, and a typical 'architectural' aspect, i.e. the quality and construction of the dwelling.

---

[1] The analysis was a principal component analysis, followed by varimax rotation. A problem here was that the number of objects (areas) to be appraised was smaller than the number of aspect ratings. In view of the facts, however, that an expansion of the study with more residential areas from similar towns would greatly add to the complexity of the study and that our study was meant to have a preliminary character, a factor analysis was considered useful.

Table 4. RESULTS OF FACTOR ANALYSIS FOR 36 ASPECT RATINGS ('housing preference scale' being excluded)

Scales		Factors		
Number	Shortened indications	D1	D2	D3
2	Green zones (imaginative-unimaginative)	-.42	.00	+.88
3	Sufficient green (sufficient-insufficient)	.00	.00	+.96
4	Appearance green (abundant-sparse)	.00	.00	+.96
5	Spatial design (spacious-cramped)	-.84	+.50	.00
6	Position of houses (imaginative-unimaginative)	-.87	.00	.00
7	Road construction (great variety-little variety)	-.84	+.44	.00
8	Street plan (clear-not clear)	-.44	+.74	+.33
9	Traffic nuisance (little-much)	-.52	+.84	.00
10	Traffic safety (safe-unsafe)	-.51	+.83	.00
11	Children's safety (safe-unsafe)	.00	+.92	.00
12	Children's play (pleasant-unpleasant)	-.40	+.74	+.33
13	Indoor noise annoyance (little-much	-.53	+.83	.00
14	Problem neighbors (negligible-considerable)	-.81	.00	+.40
15	Neighbor noise annoyance (little-much)	-.91	.00	+.34
16	Privacy invasion (little-much)	-.92	-.33	.00
17	Privacy (very much-not at all)	-.78	+.46	+.39
18	Private space around house (much-little)	-.83	+.37	+.40
19	Visual impression (tasteful-ugly)	-.74	.00	+.62
20	Representation of nature (considerable-not at all)	.00	.00	+.94
21	Street liveliness (lively-dull)	.00	-.93	.00
22	Shopping-daily necessities (sufficient-insufficient)	.00	-.80	.00
23	Shopping-goods of lasting quality (sufficient-insufficient)	.00	-.91	-.40
24	Design area (varied-monotonous)	-.96	.00	.00
25	Outdoor recreational facilities (sufficient-insufficient)	-.67	+.39	+.58
26	Impression of area (colorful-colorless)	-.75	+.40	+.49
27	Pleasantness of area (pleasant-unpleasant)	.00	-.84	+.44
28	Social isolation (not at all-considerable)	+.36	-.70	+.51
29	Provision of schools (enough-not enough)	.00	+.69	+.36
30	Recreation facilities (enough-not enough)	.00	-.87	-.47
31	Parking space (enough-not enough)	-.74	+.65	.00
32	Distance between blocks (pleasant-unpleasant)	-.94	.00	.00
33	Age of area (pleasant-unpleasant)	-.92	+.32	.00
34	Construction of houses (luxurious-simple)	-.82	+.48	.00
35	Quality of houses (good-bad)	-.85	+.46	.00
36	Value of houses (expensive-cheap)	-.83	+.47	.00
37	Familiarity with area (know very well-not at all	.00	-.88	.00
	K^2	14.00	12.00	7.20
	%	39,4	39,2	19,6
		Total	91,2%	

Privacy is closely related to both of these aspects. For instance, the better the construction, the better is the noise insulation (auditory privacy).

A spacious, varied lay-out of a residential area expressed in a network of various types of roads and an imaginative position of the houses in relation to each other will probably add to feelings of territorial privacy, for instance, by enabling the residents to become familiar with or to identify with pieces of land which are 'their own.' These are just assumptions, however, which have to be verified or disproved in further studies. As far as visual privacy is concerned, this may be stimulated by a spacious planological design: narrow streets increase the possibility of invasions of visual privacy.

Since Dimension 1 consists of two aspects which from the viewpoint of interpretation must be treated separately, we have given this factor the dual label 'Comsit' (<u>com</u>fort and <u>sit</u>uation). The relation of this dimension to the housing preference score is shown in Table 2.

Dimension 2

This dimension was next in importance (D2, Table 4). It includes aspects 9, 10, 11 and 13, together composing the aspect 'traffic annoyance.' It also includes the aspect 'shopping' (aspects 22 and 23), the aspects 'pleasantness of the area' and 'opportunities for social contacts' (aspects 21, 27 and 30), and the aspect 'familiarity with the area' (aspect 37). This second dimension correlates positively with all these aspect ratings, except for the traffic aspects with which it correlates negatively. The aspect of 'familiarity' (37) may pertain to opportunities for pleasant social contacts, e.g. recreation and shopping. In view of this we suggest calling this second dimension 'Social functions' (see also Table 2). The negative relationship that we found in our study between traffic annoyance and social functioning may be ascribed to the situation prevailing in the old center of Uithoorn. This center is much frequented, has a great many shops and is dissected by a busy road. It would be interesting to see whether the traffic aspect ratings could be eliminated from factor 2 if this kind of study were repeated in a city where the aspects 'traffic' and 'social functioning' have been deliberately separated.

Dimension 3

This dimension, the least important one in the analysis, is predominantly concerned with aspects pertaining to green zones provisions. This dimension (D3, Table 4) we called 'Green zones' (see also Table 2).

The Prediction of 'housing preference'

After a statistical analysis by the Newman-Keuls procedure (5), the difference between the areas with regard to the three dimensions and to housing preference (see Table 2) proved to be significant in some cases. As far as housing preference is concerned, areas 1, 2 and 3 are less often preferred as compared with areas 5, 6 and 7 which are much more popular. Also for area 8 low preference was expressed.

The factor 'Comsit' follows housing preference closely with areas 1, 2 and 3 least preferred, as compared with areas 5, 6 and 7. There is an unmistakable difference between these two groups of districts in favor of the latter group. For the other two dimensions, appraisals do not correlate closely with housing preference. In fact, for 'Social functions' the old center proved to have obtained a much higher appreciation score than any of the 7 other districts. As to the dimension 'Green zones,' district 8 turned out to have a significantly lower score than the other districts. Moreover, district 1 scored significantly lower than districts 4, 5, 6 and 7. For the remaining districts, the differences between the various districts on this dimension were not significant. That district 8 received such a low score with regard to green zone provisions is hardly surprising in view of the fact that at the time of our study the 'green' of this area was newly planted.

Table 5. INTERCORRELATION OF DIMENSIONS WITH HOUSING PREFERENCE

	Housing preference	Comsit	Social functions	Green zones
Comsit	+.90	1.00	- .01	- .01
Social functions	-.26		1.00	.00
Green zones	+.32			1.00

In Fig. 2, the values of Table 2 have been shown graphically for each of the three dimensions in turn. The congruence between the curves for 'Comsit' and the housing preference score is clearly much more pronounced than that for the other two dimensions.

Table 5 shows the correlation between the three dimensions and housing preference. These correlations have been based on the data of Table 2. Here too, it is clear that the 'Comsit' scores correlate most closely with the housing preference scores, followed by the dimensions 'Social functions' and 'Green zones.'

The relationship between the preference for a certain residential area and the three dimensions can also be expressed in a formula by means of which the housing preference scores for the various districts of Uithoorn can be estimated on the basis of the dimension scores. To this end, use has to be made of the strength of the correlation between housing preference and each of the dimensions (Table 5). The prediction, in the form of a multiple correlation equation, will then run as follows:

Housing preference = +.90 'Comsit' -.26 'Social functions' +.31 'Green zones'

Table 6 gives the results of this formula, if applied to the dimension scores for the various residential areas.

From this table it can be seen that the pattern of the housing preference scores corresponds very closely with the pattern of the formula-predicted scores. The correlation between the results of this calculation and the housing preference scores is very high indeed (+.97). Because averages of scores have been correlated, the above mentioned correlation coefficient has to be interpreted very cautiously. Nevertheless, even if we use it as a tentative measure of the strength of the relationship between observed and estimated housing preference scores, we may safely assume that most of the variance of the housing preference score can be explained by the three dimensions. This means that, at least as far as the present study is concerned, the hypothesis that housing preference is determined by only three general aspects of the residential area has been confirmed. It goes without saying that the value of this approach will have to be tested on new material, i.e. by means of other groups of subjects and for areas in other municipalities.

Table 6. OBSERVED AND PREDICTED HOUSING PREFERENCE SCORES (standard scores)

Areas	1	2	3	4	5	6	7	8
Observed	-1.28	-0.91	-0.99	+0.33	+0.75	+1.22	+1.44	-0.56
Predicted	-1.27	-0.92	-1.01	+0.32	+0.88	+1.38	+1.15	-0.52

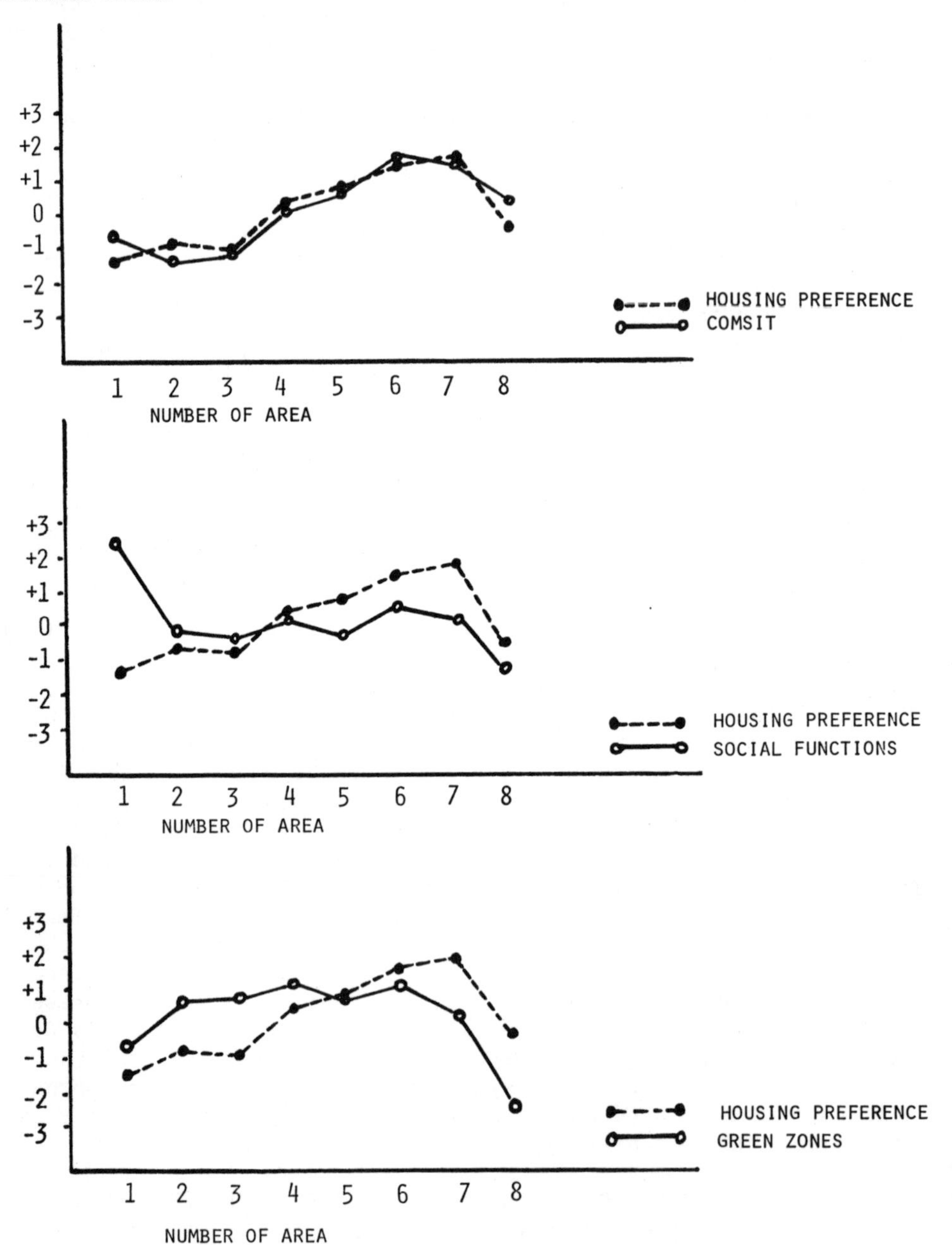

Fig. 2. Graphical representation of dimension scores of residential areas for the 3 dimensions and of scores for housing preference.

DISCUSSION

Our data show that there is a very strong preference for housing which rated high in two basic aspects, comfort and privacy, which we have combined in the dimension "comsit." By contrast, the dimension which we have called "social functions" correlated negatively with housing preference. It appears that a strong wish to live in a certain neighborhood is inversely proportional to the liveliness of that area. This includes such aspects of social functioning as recreational opportunities and shopping facilities. Apparently these interfere with perceived privacy needs. It is significant that the old center was appreciated for its social amenities but not preferred as a place to live.

What are the conclusions that might be drawn from this data, assuming it accurately reflects the desires of a typical population for a given new housing development? It appears that the desirability of housing will increase if the road network and other aspects of site planning avoid the appearance of monotony, and create a feeling of spaciousness, particularly that which fosters privacy. As might be expected, the perceived quality of construction of the house itself increases its desirability, but perhaps even more significant is the sense of a private territory, from both visual and an auditory point of view. Traffic noise and safety does not appear to be significant in preference for a housing area, if is does not intrude into the private territory of the home. The finding that the presence of shops, restaurants, etc. in the old center lowered the desirability of that area as a place to live suggests an unexpected point in favor of so-called "dormitory cities," at least for the population studied. Of course, our study only concerned activities outwardly evident at the time, and there may well be other dimensions of sociability and liveliness that would be viewed more positively in terms of permanent residence.

By contrast, this study indicates that green space as such has little influence on the attractiveness of an area for residence. It may be, however, that our respondents were influenced by the fact that all the study areas were relatively well provided in this respect. In any case, all conclusions drawn here pertain only to relatively small and rapidly expanding municipalities which are situated next to large cities. To what extent this may apply to other types of communities will have to be verified by other research.

A major research question remains to be settled in determining the utility of this approach in the practical planning of residential communities. That is the degree to which objective factors, which appear to give rise to the subjective appraisals we studied, can be found to predict satisfaction in advance of construction. Quantification of green space proved to be relatively simple in this study.

A simple procedure was developed to overlay color photos of the subject areas with a grid of small scales, from which the percentage of green was readily calculated. A product-moment of +.54 was found between percentage of green zones taken at random and the dimension scores for these areas. The relationship proved to be non-linear to a certain degree. When a logarithmic transformation was applied, a product-moment correlation of +.80 was found.

However, attempts to find measurable objective correlates for the other appraisal aspects was unsuccessful because of the great complexity of the physical and social variables. But if we can manage to find specific characteristics which strongly correlate with preference, we should be able to predict housing satisfaction for a particular area on the part of a particular population before construction begins. Such predictors, if they are found to be accurate, would of course be an important tool for designing residential areas.

REFERENCES

1. BPUR. *Building Performance*. London: Applied Science Pub., 1972.

2. Franke, J. and Bortz, J. *Beitrage zur Anwendung der Psychologie auf den Städtebau. Vorüberlegungen und erste Erkundigungsuntersuchung zur Beziehung zwischen Siedlungsgestaltung und erleben der Wohnumgebung.* Zeitschrift für experimentelle und angewandte Psychologie, 9, 76-109, 1973.

3. Manning, P. *Office Design: A Study of Environment*. Liverpool: Pilkington Research Unit, 1965.

4. Peterson, G. L. *A model of Preference and Quantitative Analysis of the Perception of the Visual Appearance of Residential Neighborhoods*. Journal of Regional Science, 7, nr. I, 1967.

5. Spitz, J. C. *Statistiek voor psychologen, pedagogen en sociologen*. Amersterdam: Noord-Hollandsche Uitgevessmaatschappij, 1965.

6. Wools, R. M. *The Subjective Appraisal of Buildings*. University of Strathclyde, Unpublished Ph.D. Thesis, 1971.

SECTION 5: URBAN SCALE DESIGN

INTRODUCTORY NOTES

Despite our emphasis on the interrelationships between community, privacy and communality, much of the discussion in this volume has focused on the private and small group end of the continuum. This appears to be true for man-environment research in general, and also for most contemporary attempts by designers and planners to create a more humane scale in our built environment. Without doubt, this is a reaction to bigness, standardization and the impersonality of mass production and mass communication which has overtaken most of the developed Western world in the past decade, as expressed in the "small is beautiful" movement. It is also a part of the widespread hunger for a sense of community in place of feeling of alienation so characteristic of urbanized people in our time. The view of community in this respect is largely one of village-scale relationships, where individuals can relate to each other personally, where they can participate in activities that affect their lives and perceive some sort of control over their environment.

This, of course, is the social scale on which the human race evolved over most of its history. There seems to be little question that in very basic ways we are biologically and psychologically programmed for that scale, as suggested by our Introduction and the opening chapter by Esser. Thus all efforts to find ways of rediscovering that scale in the built environment of industrialized peoples can only be welcomed and must be considered essential to our survival as social beings. On the other hand, the evolution of culture in the relatively brief part of human history since the agricultural revolution and the explosive growth of populations and cities in the even briefer period since the industrial revolution strongly suggests that this scale is not, in DeLong's terminology, an absolute *a priori* "good."

Much of what we call civilized social behavior cannot be confined to such a scale. Certainly, the modern technological environment cannot be managed only on that scale. Again, we return to the matter of context. Human beings as emotional mammals retain their need for village-sized social identity. But human beings as conceptualizing, intellectually exploratory, technic beings cannot function fully on that level. The chapters in this section address the more public side of the equation, looking at communality not as a close association of intimate primary groups, but in the classic sense of the *polis*. This provides one possible answer to the quest for a design for communality, raised in the first chapter: development of a model for a community of diversity.

In the history of urbanization, the center of identity for

city-as-community has been in those places set aside for the physical exchange of goods and services and conceptual exchange of information and ideas. These also set the larger scale boundaries of the collective personality of the true public citizen, even while they combine, in hierarchical form, village scale "neighborhoods" for private or personal associations. While the marketplace of commodities and the marketplace of ideas have been complementary from very early times, increasingly it is the exchange of information, rather than exchange of resources directly, that enables most of us to deal with our environment. Information itself becomes the primary resource, mediating accessibility to all others, and regulation of the boundaries that separate individuals or groups from sources of knowledge become crucial both to privacy and communality. The sharing of information is, indeed, a form of communality in itself, and we speak of the "scientific community."

The first chapter in this section by Chase considers the relationship of the built environment to the public community of knowledge by focusing on that most truly urban of institutions, the museum. Chase views museums as learning environments and compares them with schools, treating information from an ecological perspective and learning as a special class of social behaviors. He notes that the development of inexpensive information technology allows each individual to maintain his own information delivery system separate from the social life of a community. As a consequence, communication no longer functions to define and sustain community life. By contrast, both schools and museums offer social environments for learning. However, the school is typically organized quite rigidly, both as to physical layout and program, whereas the museum allows freedom in which both behavior and space can be improvised. In contrast to the school, with its fixed conceptual and physical boundaries and a well-defined population, the users of a museum are a constantly changing mass of people who do not perceive themselves as part of a group.

Chase notes that the school has been remarkably successful in transmitting specialized information, whereas museums are more satisfactory for stimulating opportunities for general understanding. But he notes that the flow of information in both cases tends to be one-way: in the case of the school, from the specialist teacher to the passive student, and in the case of the museum, from the objects themselves to the passive viewer. Here, again, we are back to the theme of contexts. One of the contexts Chase examines is that of the rules that regulate behavior, which he feels can be just as influential as the physical structures in which they are applied. He advocates changing the rules to permit greater participation by the museum users, who should be able to modify the museum environment to suit their personal histories, needs and interests. Perhaps most importantly, from the point of view of this volume, he advocates structuring the museum environment to allow individual users to form spontaneous groups in which they share experience and information. This self-organizing behavior can be seen as part of the synergic approach described in the first chapter by Esser.

In terms of the different modes of experience, defined by Hall as *high context, low information,* on the one hand, and *low context, high information* on the other (1). Chase finds that schools typically foster the latter kind, and museums the former, and he wishes to expand the contextual aspects of both by altering the rules of use, i.e., the behaviors. In the final chapter Greenbie relates such experiential situations to the physical structure of large-scale urban environments. He also relates these to the hierarchical structure of the human brain, with its different data processing capacities, and suggests that the different cognitive worlds involved not only have differing functional meanings, but very different spatial dimensions.

In the Greenbie view, there are two kinds and levels of privacy and communality. One is the continuum between the privacy of the individual and the communality of the small primary group. It is on this level that the more primitive, territorial impulses and needs are most likely to appear. In human societies these vary in form of expression by culture; they are the kind of behavior-space situations which Hall called *proxemic*. The other level concerns the privacy of small groups relative to other small groups in culturally diverse larger environments. Behavioral situations on this level will tend to be higher in information and lower in context than proxemic ones. To emphasize the experiential difference and particularly the *cross-cultural* nature of this second level of privacy and communality, Greenbie uses the word *distemics*.

Chase's museums are an excellent example of distemic relationships. Much of the kind of learning provided in them would be entirely unnecessary if users were of a single proxemic culture. As Chase notes, people go to museums, not as members of a group, but as individuals, seeking some larger communality that their private lives cannot provide. The formation of spontaneous groups formed around the sharing of information by social strangers in short-term contexts is the essential function of the large-scale distemic social environment. It is, indeed, what urban life is all about. It is this kind of communality which is endangered most by the social fragmentation of modern life. The proxemic/distemic continuum is illustrated by Greenbie with examples from sports and the structure of an ethnic community, physically bounded within, but closely related to, the communality of a central city. The final argument is that it is the role of the designer to unite all these dichotomies on a flexible continuum within the universal context of human experience.

1. Hall, E. T. *Beyond Culture*. Garden City, N.Y.: Doubleday, 1976.

THE SOCIAL ECOLOGY OF THE MUSEUM LEARNING ENVIRONMENT: IMPLICATIONS FOR ENVIRONMENTAL DESIGN AND MANAGEMENT[1]

Richard Allen Chase

Department of Psychiatry and Behavioral Sciences
The Johns Hopkins University School of Medicine
Baltimore, Maryland 21205

COMMUNITY, PRIVACY, THE QUALITY OF EXPERIENCE: CASE STUDIES FROM MUSEUM LEARNING ENVIRONMENTS

This paper contributes to the Workshop on Community and Privacy from the vantage point of a single social institution: The museum. During a time when so many social institutions seem to be failing in their efforts to meet the needs of the public they serve, many museums are growing and differentiating in a manner that is making them more useful and more interesting to increasing numbers of people. Study of the adaptive changes that are occurring in many contemporary museum learning environments reminds us of the important opportunities for public institutions, properly designed and managed, to enhance the quality of private life.

MUSEUMS AND LEARNING

At a time when so much of our formal system of education disappoints us, museums seem to be gaining increasing public attention and use (15, 28, 35, 39). The recent widespread growth of museum education staffs and education programs suggests that museum professionals and the public they serve are finding encouragement in

[1] A major portion of this paper is based on a lecture delivered as part of the Exhibits/Programs Workshop of the Association of Science-Technology Centers, Center of Science and Industry, Columbus, Ohio, November 15, 1974. These portions of the text have been previously published in *Museum News* under the title: "Information Ecology and the Museum Learning Environment."

the view that museums have great potential as learning environments. It is the purpose of this paper to examine this potential in some detail.

The Evolution of Museums as Learning Environments

What makes the museum different from schools, theaters, and other learning environments? Museums are notably filled with objects. We have assumed, largely on faith, that important objects can teach us important things. This faith still seems valid to most of us. However, the early stages in the evolution of museum learning environments was characterized by a disproportionate concern about the objects themselves. More recently, interest has grown in the question of how objects actually do teach. This interest requires a closer look at the ways in which people relate to objects and the ways in which behavior is changed through this relationship. This interest requires us to think about the museum environment as a whole; the ways in which physical structure and social structure influence information seeking and information using behaviors; and the ways in which the special past history of museums allows special opportunities for learning experiences to occur.

Museums and Schools

We begin our evaluation of museum learning environments by contrasting them with the more numerous and better known learning environments of schools. The student comes to the school for a specific purpose, and the school outlines in some detail the responsibilities and costs that must be met to achieve that purpose. School learning experiences are organized according to plans, and these plans are translated into courses and programs of study that are similar for every student who elects them. Courses and programs are mapped onto the physical spaces of the school -- which have been designed for relatively rigid use patterns.

In contrast, most museums allow free access, and do not require that the user enter into a formal contract before the environment can be used. You don't have to make your purpose known before you can enter a museum. As a matter of fact, you don't even have to be aware of the purpose yourself. Once in the museum, we are impressed with the design of the physical spaces and the objects that fill them. A sense of excitement about the building and its contents is commonplace, and these feelings are amplified by the knowledge that there is no rigid plan for how the environment is to be used. Most of the space is available on a free-access basis, and trails can be freely improvised. Continued use of the same museum will reveal that the physical environment changes more often than is the case for most schools. In contrast to the groups in school, the users of the museum are part of a constantly changing mass of people, and there is little perception of being a member of a group.

Museums and General Education

Our schools have become well adapted to support education in the specialist tradition. A major influence behind the growth in the number and size of schools is the growth of new fields of enquiry, and the continuing differentiation within existing fields of enquiry. Within this tradition, formal contracts between school and student and rigid plans for organizing learning experiences and the physical spaces that support these experiences are adaptive. The machinery of schools has supported an almost unimagined productivity in the generation of specialized knowledge. However, this vast and growing inventory of specialized knowledge is fragmented and dispersed across a multiplicity of fields. Is it possible that museums can build upon their strong traditions of broad interests, and informal, flexible, and democratic character, to enhance opportunities for general education in an age that so desperately needs broad understanding of the problems and opportunities that confront human populations (14, 19)? Can museums build upon their strong commitments to general education in a manner that will help us achieve a sense of wholeness in our understanding of human experience? This is a bold vision, but it addresses urgent needs and responsibilities. Schools have, for the most part, succeeded better in their support of specialty education than they have in their support of general education (26). Perhaps schools won't be able to perform a significant integrative function during this time of social change. Perhaps this role will fall to different classes of learning environments (5). If this is the case, then museums might well be on the threshold of just such important new responsibilities and opportunities within an evolving ecology of learning environments (37, 40).

THE LEARNING PROCESS/ORGANISM-ENVIRONMENT INTERACTION

Psychophysiology of Learning

Curiosity, exploration, and play are important components of the human behavioral repertoire starting with infancy, and continuing throughout the entire life cycle (22, 24, 25, 30, 32). Infants show great interest in object quality, the use of objects as tools, the use of objects as mediators of social interaction, and the way things work. Infants in the age range 9-18 months spend almost half of their time looking at and playing with physical objects (12). During the first three months of life, infants relate to objects mainly by looking, holding, and mouthing. From three to six months they add hitting, shaking, and examining. Pulling, tearing, rubbing, squeezing, sliding, pushing, dropping, and throwing are observed to appear in infants six to nine months old. Socially-instigated behaviors, such as showing and naming, are observed toward the end of the first year (38). From these, and many other observations, it is clear that the human nervous system is organized in a manner that predisposes us to show interest in novelty and change in our environment, and equips us with a broad range of exploratory behaviors (30).

Learning/A Social Behavior

The exploration of objects, events, and even ideas and feelings does not occur in environments that are lacking in opportunities for, and encouragement of, exploration (7, 16, 17, 18, 21, 31, 34). For these reasons, we can consider learning behaviors to be a special class of social behaviors. Human infants must first achieve a successful psychological and social attachment to competent adult caretakers before they can venture forth to explore the surrounding environment (1). This fundamental fact of human biology helps to remind us that learning and sociality are inextricably inter-related in our life careers.

Making Contact with the Existing Behavioral Repertoire

Learning involves growth and differentiation of an already-existing behavioral repertoire. Learning requires experiences that are compatible with existing interests, needs, and skills. In order to be effective, learning environments must make contact with the past history of the organism (4). The designer of a learning environment must therefore know a good deal about the past history of those who will be making use of the new environment. This allows the design of an environment that will, from the outset, make sense to its users, be capable of generating interest, support exploration, and provide opportunities for new behaviors to be acquired.

These points can be clarified through an example. The Elizabeth Randall Junior Museum in San Francisco operates a continuing project that allows children to design, build, and use their own go-carts (Figure 1). This is an attractive project for pre-adolescent children, largely because of the glamour and excitement that is attached to wearing a helmet and goggles and speeding across the parking lot

Fig. 1. This boy has built his own go-cart at the Elizabeth Randall Junior Museum in San Francisco.

and roadway system around the museum. In this case, the whole sequence of learning experiences begins by making possible an achievement that corresponds to the psychological needs and interests of the children. The pursuit of that possibility spans months of time. These sustained learning behaviors are supported by the social reinforcement of teacher and peers, and the visible emergence of a piece of machinery that will inevitably elicit additional admiration and respect.

Learning/The Sense of Wonder

"Wonder is the foundation of all philosophy."
[Montaigne, Essays, Bk. iii, Ch. 11, 1595].

When we enter a museum and encounter a reconstructed whale or a dinosaur skeleton, most of us do feel a sense of wonder. Whether this emotional energy gets put to more concentrated and focused use, as required for learning, depends on how the learning environment is designed and managed. For the most part, our museums do a fine job of evoking wonder; on the whole, they seem to be better at this than most schools. But when it comes to concentrating and focusing on specific questions and ways of answering them, most museum learning environments do rather less well than schools. Here is an instance in which the museum might well borrow some lessons from the school, and address the general question of how the wonder, excitement, and curiosity that good exhibits evoke can be channeled into learning how to ask good questions and find answers to them.

The Importance of Consequences

What do you think will happen if you hand a 6-month old infant a disc of shiny metallized mylar plastic? The results are really quite striking -- the infant grasps the thin plastic sheet, and continues to crumple, rub, poke, twist, and tear at it for ten or fifteen minutes with sustained interest and enjoyment (9, 11). We might ask how such a simple object can elicit so much behavior from such an inexperienced person. We do know things about the behavior of infants that help to provide an answer to this question. At six months, most infants have just achieved the capacity to use their arms, wrists, hands, and fingers in the ways we have described above and we know that when behaviors first occur in the course of development they are used a great deal, and tend to overshadow some of the older and more familiar behaviors (38). We also know that infants are interested in and pleased with sharp, high-frequency sounds such as the crinkling sounds produced by twisting the mylar sheets. Infants also show interest in bright lights, and particularly in changing patterns of bright lights. The shiny metallized mylar sheets reflect light in just this way. And so we come to understand that a metallized mylar sheet is ideally suited to the motor capacities of a 6-month old infant, and the consequences of its manipulation produce visual and auditory consequences of a sort that are of great interest. When we consider why it is that such young infants play with the object for so long without adult encouragement, we have a chance to learn about another behavior principle: behavior is sustained by positive and interesting consequences (33, 36).

Of all the principles we consider in the design of learning environments, this is probably the most important. Wonder, excitement, and curiosity can elicit behavior, but there must be an opportunity for that behavior to have consequences that are of interest if the behavior is to be sustained. This principle is in operation in the case of the 6-month old infant playing with a disc of metallized mylar plastic. It is also in operation in the case of the children in San Francisco who are building go-carts. A major limitation of many museum learning environments is inadequate provision of opportunities for the learner to behave. There are few opportunities to put new knowledge to use in ways that can have immediate, interesting, and important consequences.

Diversity of Interests and Temperament

We have discussed the importance of making contact with the past history and existing behavior repertoire; the importance of wonder, excitement, and curiosity; and the important role of consequences for the success of learning experiences. When someone comes to a museum or a school, what do we know about the past history of that person, and the kinds of experiences most likely to elicit wonder and curiosity? We usually leave it to individuals to explore the available learning opportunities and make their own selection from among a large number of possibilities. There are some obvious problems with this. Fields of knowledge are numerous and complex, and the brief descriptons of them in course catalogs often fail to provide a meaningful understanding of an available learning opportunity. In the case of the museum, the building and its collections are, in a sense, a vast three-dimensional catalog of learning opportunities. As in the case of fairs and markets, the exploration of the environment itself can consume enormous amounts of time and overshadow the opportunity to focus attention and interest in a way that predisposes to effective learning. Coupling diversity of choice with effective opportunities for concentrating attention and effort defines one of the special problems for the environmental designer working the museum. We are going to return to this question later when we talk about the design of exhibits. At this point, it might be helpful to suggest that the learning environment, whether museum or school, might profitably invest in an effort to obtain a history from each user, and explain the learning opportunities that correspond most closely to the needs, interests, and even the temperament and learning style of the user. Computers with self-access terminals could be used for this purpose. The computer could make its suggestions in writing, and the document produced could serve as an individualized catalog, plan, and guide. Individual profiles of interest areas and learning-styles, used in these ways, could profoundly influence the relationship between users and the museum environment. The profile data could be used to generate a broad range of other individually-responsive documents such as worksheets, notices of relevant future events, the matching of users with similar interests, and so forth.

LEARNING ENVIRONMENTS/INFORMATION

Information Ecology

It is useful for those who design learning environments to think about information in much the same way that an ecologist studying the life of an animal population in nature would think about food (10). After all, the users of learning environments are in search of information. From an ecological perspective, we think about learning environments in terms of the variety, amount, and quality of information they contain; the information needs and communication behaviors of user groups; and the physical and social design features of the environment that influence the ways in which people and information are related.

Contingency Management

We have talked a good deal about the importance of matching the information in a learning environment with the past and present needs and interests of students. We have not yet spoken about the organization of sequences of behavior in ways that facilitate learning (33). Some of the rules that are used in school have been designed to deal with this issue. When you read through a university catalog you see that certain courses must be successfully completed before the next course in a sequence of study can be taken. Admission to a more advanced course in a sequence of studies is made contingent upon successful completion of the more elementary courses that have been designed to precede it. The museum environment rarely discriminates between entry-level and advanced-level learning experiences, and almost never requires that the first be completed before access to the second is made possible. However, in some circumstances it can be very helpful if students have such a structured experience to guide them through complex subjects. Movement from one sub-environment of the museum to another could be made contingent upon a preceding successful learning experience in much the same way as progression through a series of increasingly advanced courses is arranged in a school. The success of learning experiences could be evaluated through problem-solving exercises requiring the use of new knowledge and skills. This process could be automated, and the document that records successful completion of one stage of a learning experience sequence could function as a ticket to the next experience in the sequence. The Bay Area Rapid Transit System in San Francisco (BART) uses hardware that is quite adaptable to the uses I am describing (Figure 2). The machine issues a ticket contingent upon depositing appropriate amounts of money, and the automatic gating system operates when the coded ticket is electronically read at the gate. Obtaining the ticket is contingent

Fig. 2a. BART tickets are sold by these vending machines which dispense a credit-card size, magnetically coded ticket usable for a single ride, or again and again until its value is used up. (Courtesy of the U.S. Department of Transportation and the Bay Area Rapid Transit District).

Fig. 2b. The magnetically coded BART ticket is "read" by machine at the gate. If the value of the ticket is sufficient, the gate opens and the ticket is returned. The ticket is also "read" by machine at the exit gate. If the value of the ticket was exactly that required for the trip, the ticket is retained at the exit gate. Otherwise, the ticket is returned. (Courtesy of the U.S. Department of Transportation and the Bay Area Rapid Transit District)

upon the expenditure of money, and the duration of the trip is contingent upon the fare paid. The contingencies are controlled entirely by machine.

I am not suggesting that all exploration through museums should be sequential, hierarchically-organized experiences contingent upon machine administered and evaluated tests. But I am pointing out that contingency management can be a very powerful determinant of learning effectiveness (29). It would be interesting to organize some parts of the museum environment for those who would like to participate in learning experiences organized in this way.

The use of a technology comparable to the BART ticket system would not only provide a way of organizing the exploratory behaviors of users in a manner that is contingent upon effective learning, but it could also provide an automated cumulative record of individual use patterns of the museum learning environment that could assist the ongoing monitoring of the quality of that learning environment. The same data could also be used to generate individual accomplishment profiles that could provide a basis for issuing diplomas, achievement certificates, and other credentials that could qualify students for teaching responsibilities and other privileges that facilitate continued growth of knowledge and skill. The structuring of contingent relationships can influence learning behaviors far more effectively than manipulation of the physical environment alone. Descriptions of contingency management systems offend some people because they tend to sound impersonal and mechanical. However, such systems can bring enormous economy, efficiency, and satisfaction to the relationship between a student and a learning environment.

Information Sources

From the vantage point of information ecology, it is possible to observe a number of potential sources of information that are rarely used in museum exhibit design. For example, in the United States and Great Britain, there are excellent low-cost kits that provide an introduction to basic topics in science, such as electricity, magnetism, light, color, mechanics, and so forth. These kits consist of simple and effective reading material, outlines for simple experiments, and the basic materials needed to perform these experiments (Figure 3). Each kit is, in effect, a small learning environment. The care with which these kits have been prepared permits amplification of the elements to a scale and character suitable for use as a full-scale exhibit in a museum. The same approach could be used to adapt the excellent programs and materials that have been developed for use in science study programs in the elementary schools by the American Association for the Advancement of Science, the

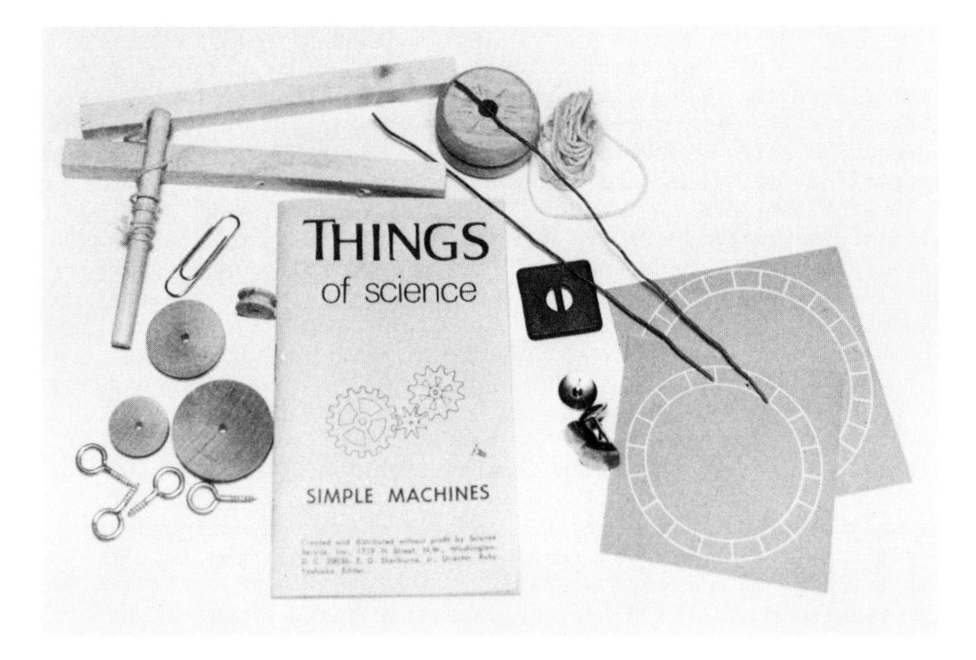

Fig 3. One of the low-cost science kits available in the United States by mail subscription from Science Service in Washington, D. C.

Education Development Center, and other groups (3). Translation of information from book or kit scale to environmental scale could result in a major expansion of high-quality museum exhibit and program experiences in science. The books and kits used for such purposes could be available in the museum shops for individual purchase and use, thereby reinforcing and extending the museum learning experience. The original kits and books could also be made available in museum media libraries for use in protected work-study areas in the museum, or for use at home.

One of the great ironies illuminated by an ecological analysis of learning environments concerns the quality and amounts of information available from students themselves. When you look about a large lecture hall at an attentive audience listening to an authority on a subject of interest to the group, you are looking at a large community of individuals which often possesses, as a group, more information about the subject being discussed than is available to the lecturer. However, the rules of the occasion prevent the members of the group from sharing information directly with one another. The same observation applies in the museum learning environment. In this case also, custom prevents the direct sharing of information between members of the group making use of the museum. It would be interesting to know how many of the users have valuable information that could add enormously to the information density of the museum learning environment if new conventions were designed to allow such information to be made explicit. Imagine, for example, small stalls like those used at neighborhood fairs and festivals and flea markets located in the museum for use by the visitors on a first-come, first-served basis. The stalls could be made available to those who want to share experience and personal knowledge consistent with the larger objectives of the museum. The science fairs designed for young people provide a model, albeit a time-limited one, for the way in which such a system might operate (Figure 4). This is but one of many plans that might be used to allow the users of museums to become contributors to, as well as consumers of, the museum's information resources. Another plan might allow users to join in small groups to share problem-solving experiences. For example, a group might be provided with a stop-watch, a length of string, a lead weight, a tack, a yardstick and a pencil along with minimal instructions about how to use these materials to build a pendulum. Their job would then be the development of a way to describe pendular motion.

There are other ways in which the information distribution systems of the museum could be improved. In recent years, museum shops have grown in size and number, thereby allowing an important new means for the dispersion of learning materials. In addition to the conventional museum shop configuration, small satellite shops could be distributed throughout the museum so that learning materials relevant to sub-regions of the museum learning environment could be directly linked to them in space and time. It would even be possible to use vending machines for this purpose. A vending machine might, for example, dispense inexpensive visual illusion cards as an integral part of an exhibit on visual perception (Figure 5).

The time and space linkages in the design of learning environments are of great importance. Objects that allow tangible, direct, personal, and complementary experiences, like the visual illusion experiment materials, are most meaningful when they are linked to other components of the museum learning environment. It is in such places that the objects become most meaningful, and it is at such times that they are most wanted. These principles have been grasped by the department store manager who links the preparation and sampling of food to the display of relevant food and food preparation products that are available for purchase.

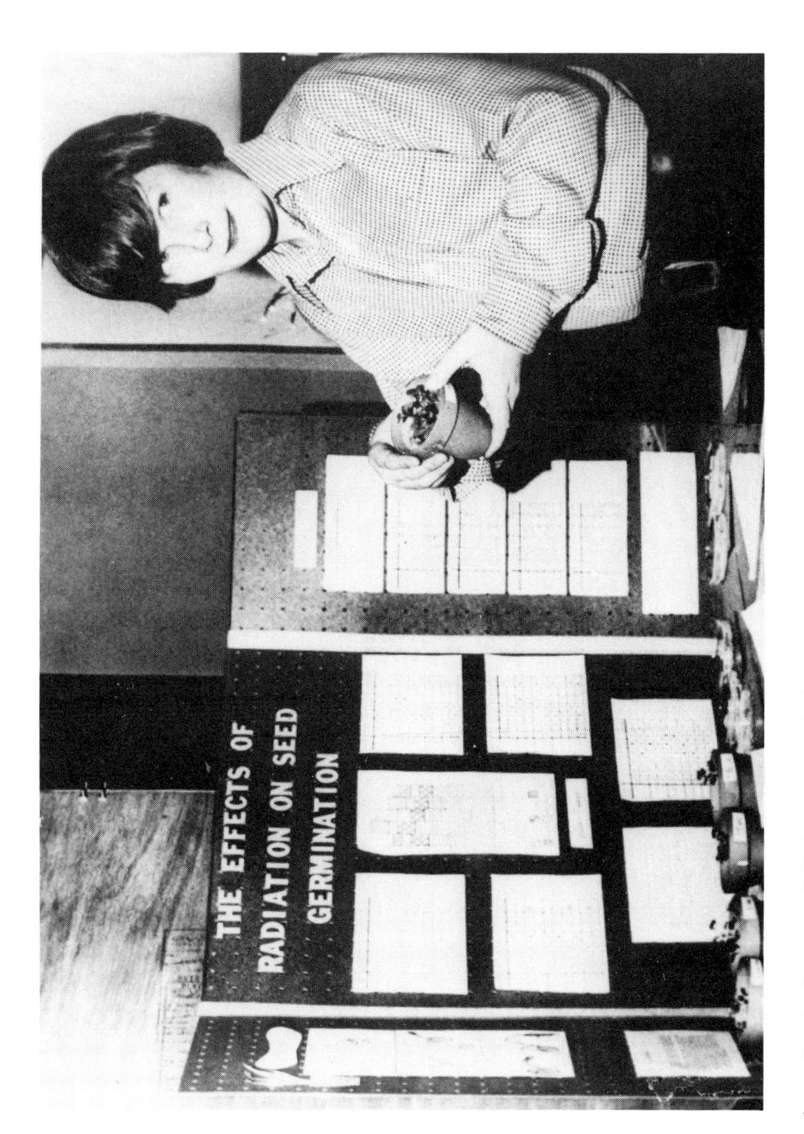

Fig. 4. A student exhibit at a Science Fair in Baltimore, Maryland. (Courtesy of the Baltimore News American. Photograph by Richard Tomlinson).

Fig. 5. What are the features of this pattern that cause us to perceive parts of it to be in motion? This, and many other patterns that result in illusions of motion, size, and color could be dispensed by vending machines in museums.

LEARNING ENVIRONMENTS/SOCIAL STRUCTURE

People, Information, and Rules

We have discussed how rare it is for those who make use of schools to share information and work collaboratively. The same can be said for those who make use of museums. In the case of the school, the passivity of the individual student is rationalized by attaching exaggerated importance to the role of the teacher. In the case of the museum, the passivity of the student is rationalized by attaching exaggerated importance to the role of the physical environment. In both cases, we are looking at customs and traditions that have been shaped by long and complex histories involving many people. However, it is possible to change customs by designing new rules.

We have already noted that learning environments can be conceptualized in terms of information resources, people, and the rules that govern the relationships between the two. The fact that many of the rules that operate in the natural ecology of learning environments have been shaped by complex histories, and have not been designed in an explicit and deliberate way, can obscure the equally important fact that such rules can be designed in an explicit and deliberate way with great economy of effort, and with great effect.

In 1971, I shared in a course exercise designed to allow university students to share information with each other (10). The name of the course was: "Environmental Analysis: Human and Social Factors." One hundred and ninety students were enrolled. They attended lectures to two fifty minute periods each week. We changed the rules for one of the fifty-minute class periods in a number of ways. Instead of meeting in a large fixed-seat lecture hall, the class was asked to meet in a large room in the student union building. Each student came to this class prepared to teach something to other students. This required prior written specification of learning objectives, ways of achieving these objectives, and preparation of any special materials that would be needed. Figure 6 shows students teaching other students during this special class exercise. Students taught each other skills of many kinds, including the use of musical instruments, drawing instruments, sports equipment, and hand tools, and they transmitted information through simple participant-observer demonstrations (Figure 7). Instead of one teacher sharing information with 190 passive students, this exercise allowed 190 student-teachers to share information with each other.

There are many questions that can be raised concerning the usefulness and appropriateness of techniques such as this. For the present purposes, however, we need only consider that the striking changes brought about in the behavior of this group were managed by simple changes in rules (23). We didn't change the student population, and we didn't change the information resources that were available to the community -- we simply changed the rules that define the relationships between the two.

The ways in which people share information through designed interactions can be a powerful facilitator of learning. Enormous attention is given to the design of the physical aspects of the museum learning environment, and, for the most part, little attention is given to the explicit design of rules that will determine the ways in which people and information will be related. A redress of this imbalance could have profound consequences for the effectiveness of the museum learning environment.

Fig. 6. Students teaching other students as part of a specially-designed exercise at Cornell University.

Fig. 7a. One of the Cornell students teaching another student how to play the flute.

Fig. 7b. One of the Cornell students teaching another student how to knit.

An exhibit dealing with mechanics could include an activity area in which self-selected teams work collaboratively on the design of simple machines. Different teams might compete with each other for the best solution to a problem through appropriate machine design. The activity area could be provided with a gallery, and the spectators who elect to sit in the gallery might be asked to provide criticisms of the performances that they observe. They might be the jury that decides on the quality of the problem-solving and design efforts they are observing. Each of these strategies implies a rule or small set of rules. The explicit use of appropriately authoritative rules allows experiences such as these to be designed just as deliberately as the physical environment of the museum is designed.

Social roles identify the specialized functions that are important to the life of a group (23). Roles can also facilitate the integration of members into a group. For example, a great many tasks must be performed to maintain an orderly and efficient operation of the network of youth hostels that are so popular in Western Europe. Within each hostel there are needs for food preparation and serving, and for a wide range of clean-up activities. Although the director of the hostel may have some regular assistants, it is often necessary to get help from the young people staying at the hostel. The process is facilitated by assigning specific tasks to individuals as they check in. This is not only an equitable and efficient way of seeing that the continuing needs of the facility are met, but also a very fine way of allowing each visitor to feel a part of a small and changing society in a tangible and effective way. Couldn't we allow users of the museum learning environment to volunteer for a wide range of jobs -- even on a very short term basis, such as a day or even a part of a day? A pre-arranged inventory of roles could be developed, such as guide, tutor, lecturer, demonstrator, supervisor, collaborator, and so forth. The jobs available within each role category could be described, and visitors could volunteer to fill them in a simple and direct way. This plan represents an expansion of the volunteer roles that are already available in most museums and a marked simplification of the procedures followed to fill such roles, as well as a reduction in the time jobs are held. The plan could allow a great many people coming to the museum each day to relate to the museum learning environment in a more constructive and meaningful way. This plan helps convert the passive experience of many museum users into an active relationship with the museum learning environment with an accompanying growth in opportunities for learning.

INFORMATION ECOLOGY AND THE MUSEUM LEARNING ENVIRONMENT

Identification, Transportation and Orientation

We have referred to the fact that the museum learning environment originates with the collection of valuable and important objects. Preservation and restoration were the earliest concerns of the keepers of such collections, later joined by concerns about interpretation and integration into education programs. The chronological priorities can still be perceived in the conventions that govern the design and management of most museum learning environments. The architectural conventions used in the design of museums communicate a concern for the safekeeping of the museum's contents that often overshadows even the most elementary signs of welcome. Massive stone facades with minimal fenestration and miniscule bronze plaques giving the name of the museum and announcing hours of operation constitute some of the most frequently used conventions of this type. Massive stone lions with bared teeth might be used as decoration, and, although innocuous enough, it would require a great stretch of the imagination to think that such beasts were placed there to make visitors feel welcome (Figure 8). Most of us have, of course, learned how genuinely friendly most museums really are, once you find your way inside. As with any new and potentially intimidating environment, a humane introduction by someone who is already familiar with it is the best possible way to come to know that the museum is usually a friendly and exciting place. Many of us have had such introductions, often as children guided by familiar and caring adults. Not everyone has had such an experience however, and it makes sense for those who want to enlarge the accessibility and influence of the museum learning environment to consider whether it has a meaningful identity for everyone who might want to make use of it. How many people actually know what a particular museum contains, the experiences that it can make available, how to get there, and how to get back home again? And for those who do arrive for their initial experience, how many are able to achieve the sense of comfort and belonging that only follows when a proper welcome and orientation is made available? These are basic questions that issue from an ecological orientation to the design and use of museum learning environments.

The Museum as an Open System/Making Contact with the Surrounding Environment

The tradition of protection and care of objects leaves us with the impression that the museum learning environment must be fixed in space. However, this need not be the case, and should not be if

Fig. 8. One of the stone lions used as decoration on the grounds of the Baltimore Museum of Art.

the valuable information available in museums is to be shared with the widest possible audience. A great many experiments in the dissemination of information traditionally confined to the museum have been successfully undertaken. Radio and television programs, publication and reproduction programs, and satellite museums have been incorporated into such experiments, and are now well-accepted practices. But opportunities to share the special qualities of the museum learning environment on a broader scale are great, and exciting experimentation remains to be undertaken. For example, the Ceylon Tea Centre in London contains a small museum dealing with tea. There are some basic exhibits showing tea plants and the tools used to harvest them, and a display board that asks some basic questions about tea, such as how many cups of tea are consumed in the world in an average day. When you press a button by the answer of your choice, light panels flash on to tell you whether your answer is correct, and, if not, what the correct answer is. The same environment contains two tea parlors and a sales desk for the purchase of package tea. In a similar manner, a tea museum has been incorporated into the McCormick Company spice processing plant in Baltimore, Maryland (Figure 9). It is easy to conceive of constructive synergistic relationships between such relatively unknown

Fig. 9. The tea museum and tea house at the McCormick Company factory in Baltimore. Artifacts are used to explain the cultivation and processing of tea, and the tea house serves hot tea and cookies to visitors and guests.

museum components within other environments and major museum environments. These relationships could be explored through network designs that allow the exchange of staff, materials, and visitors, and collaborative design of exhibits, activities, and programs (27).

Many parts of the natural ecology of man-made environments could be designed to function, in part, as learning environments (6, 8, 13). The giant machines that support the metabolism of human communities could be uncovered and explained. The power plant at O'Hare Field in Chicago is contained within a glass building. This is an exciting move in the direction I am speaking about because it allows visual contact. By unmasking itself, it develops a potential to arouse a sense of wonder, excitement, and curiosity -- those generators of learning behaviors which have been discussed earlier. How much better it would be if we could get even closer to the equipment, particularly if it were labeled and arranged to teach us how it works. A major new plant for the re-cycling of waste materials is being built in Baltimore, and the opportunity to use this facility to instruct the community about pollution ecology, energy, and the re-cycling of waste products is being seized to some extent. A visitor center with an exhibit and demonstration area is being built on the site, and the entire plant is being covered with bold graphic designs in yellow and orange paint (Figure 10).

The number of regions within an increasingly incomprehensible man-made environment that could be uncovered, made accessible to exploration, and designed to be informative is virtually limitless. The Xerox machine takes a coin and gives an exact replica of a printed page without a hint about how the miracle is achieved. In a similar manner, the Coke machine (even though labeled "The Real Thing") performs a somewhat more pedestrian miracle when it takes coins and gives back the drink we are after -- all shrouded in the secrecy of brightly-enameled sheet metal. Figure 11 shows what can happen when the skills of those who design museum learning environments are applied to new frontiers. This photograph shows a

Fig. 10. This is a model of a new waste-recycling plant (Pyrolysis Plant) being built in Baltimore, Maryland. A graphic designer in the Department of Planning has used this model to show the large graphic designs that are being painted on the plant in yellow and orange colors. Notice the large dots on the tank to the left, and the elaborate design based on the letters of the word "pyrolysis" painted on the side of the large building to the right.

Fig. 11. The opaque panel that covered the upper portion of this coffee vending machine has been removed and replaced with a translucent sheet of plastic. It is now possible to see how the machine works.

coffee vending machine that is part of the automatic food service in the Center of Science and Industry, Columbus, Ohio. The museum staff has replaced the opaque panel that conventionally conceals how the machine works with a translucent sheet of plastic. How much more interesting and informative the machine becomes when it is uncovered. I am told that some people drop in their coins simply to see how the machine works, and then walk away leaving a full cup of hot coffee.

There are many ways in what the membrane that surrounds the traditional museum learning environments can be made at least semi-permeable, so that information can flow freely across it. We have just considered some of the ways in which the techniques of the museum designer can be applied to regions of the surrounding natural ecology. However, these extensions of the museum environment will serve better if they function as part of a network of museum and museum-like learning environments. Maps, guide books, combination tickets, transportation systems, and symbolic incentives can facilitate the extent to which each element in a network of museum learning environments gives encouragement to the use of the others (8).

The effectiveness of the museum learning environment can also be enhanced by moving parts of the surrounding natural ecology inside. For example, a chest x-ray van brought into the museum could be examined and explained, and used as a compelling focal point for health education programs. Small sections of factories could be brought into the museum to make available for close observation those tangible and direct operations that allow us to understand the basic trades and industries. It would also be valuable to bring into the museum environment some of the actual problem-solving tasks that require the availability of high-quality information. If a town council is deliberating about pollution-control ordinances, why couldn't some of those deliberations be moved into the museum so that they could be directly observed? What better way to generate a respect for the importance of creating and protecting knowledge than seeing it put to use in important ways? Museum visitors could also share such experiences through simulation exercises modeled on town meetings, debates, jury deliberations, and elections (2).

Environmental Analysis and Environmental Management

Evaluation of the extent to which the museum serves the information and experience needs of its users requires an ecological orientation. The extent to which learning occurs, and the extent to which a particular learning environment facilitates learning is best evaluated by studying the interactions between people and their environments. Different individuals can make use of the same learning environment for very different purposes, and a narrowly-focused test or examination might miss this point entirely. On the other hand, evaluation of the effectiveness of a learning environment through close examination of its physical structure alone can be equally misleading. We need more accurate information about who comes to the museum learning environment, for what purposes, and to what ends. These questions can be addressed in a modest and descriptive way to begin with, and the resulting information used to contrast the museum learning environment with other learning environments, such as schools, libraries, and theaters. Ultimately, we will want to understand the museum as a subset of a differentiated system of learning environments. We will want to understand the functions for which it is best adapted. We will want to understand the ways in which opportunities for learning in the museum environment can be enhanced by appropriate linkages to other learning environments, as well as the ways in which learning in other environments might be enhanced through appropriate linkages to museums. This knowledge base could allow more intelligent planning and stewardship of the next stages in the evolution of museums within the larger framework of an expanding ecology of learning environments.

COMMUNITY, PRIVACY, AND THE QUALITY OF EXPERIENCE

"Community" and "privacy" are terms that suggest the importance of balancing the relationship between public life and private life. In these concluding remarks, I would like to share some of my concerns about community and privacy in contemporary social life. To my mind, we are not so much threatened by distortions in the balance between community and privacy as we are by the erosion of the quality of both communal and private life. In the advanced technological societies, the environments we live in are increasingly man-made, vast in scale and complexity, and increasingly incomprehensible and inhospitable. These characteristics undermine the quality of public life, the quality of private life, and the relationships between the two.

Features of Contemporary Environments that Predispose to Social Isolation

A number of features of contemporary social environments seem to me to force on their occupants an increasing degree of social isolation.

1. Limited shared facilities for spontaneous and intimate social interaction.

In many major American cities the migration of middle-income groups out of the city and into the suburbs has resulted in underutilization of facilities that traditionally allow a community to view all of its members, with opportunities for casual encounters to grow into friendships. These facilities include plazas, malls, parks, and promenades. As population leaves the city, and distributes itself in suburbs, the regional shopping center appears. It is developed by private entrepreneurs interested almost exclusively in the specific economic exchanges essential to the well-being of the store owners. The great social function of markets, plazas and town centers are given little attention under these new conditions. As a result, opportunities for social interaction are diminished in the city center and suburb alike.

2. Intimidation of social interaction through "defensive" and "anonymous" architectural styles.

The social instability and violence evidenced in the 1960's gave rise to increasingly defensive architectural styles. Schools without windows constitute one of the consequences of this style. Economic constraints have dictated the elimination of most of the detail from contemporary buildings, giving rise to an increasingly anonymous style in contemporary architecture. Defensiveness and anonymity symbolize inhospitality in architecture, and further contribute to discourage social interaction in and around public buildings.

3. Decrease in the size of privately-owned spaces with attendant reduction in the ability to accomodate groups within one's own home.

The increasing cost of housing in the city has resulted in a progressive decrease in the amount of space individuals are able to rent or buy. This limits the ability to collect objects, the ability to engage in hobbies, and the ability to accommodate visitors and entertain groups. Under these conditions, public accommodations are increasingly relied upon for functions that traditionally were accommodated, at least in part, in the home. Solving the problem of how one can "be at home" in public, becomes an increasingly important dimension of the environmental design of public amenities.

4. The use of space to achieve status through distancing and sharp definition of boundaries.

Those who migrate from the city to the suburb seek greater security and comfort. In most suburban settings, size of homes, and amount of space surrounding homes constitute desired evidences of high social rank. Unfortunately, these very characteristics of housing can minimize opportunities for casual social encounter and the meaningful friendships that such encounters can give rise to.

5. The development of mass media and an information technology that allows each individual to maintain his own information delivery system separate from the social life of a community.

Television, and inexpensive machines that make use of records and tapes allow individuals to have unlimited access to low-cost information. As a result, much of the information an individual now consumes is unrelated to the life of the community he is a part of. Communication is no longer dependent on community, and no longer functions to define and sustain community life.

6. Increases in the scale and complexity of social systems diminish their responsiveness to the needs of the individuals who work and live within them.

As all social organizations have grown in scale and complexity, they have generated rules that simplify the conduct of their principal operations. The result is a stereotypy and homogenization of social interaction accompanied by feelings of depersonalization and inability to influence the social environment of which one is a part. These characteristics seem to apply across a broad range of complex social institutions including the modern corporation, government agency, and university.

These features of contemporary social life portray a weakening of both community and private life. Erik Erikson has commented on these issues in his recent Jefferson Lectures:

> The main point is that at each state of life there must be a network of direct personal and communal communication safely set within the wider networks of automobility and mass media. A spontaneous readiness for such community forming is noticeable all over the country and is reaching onto the architects' drawing boards. Such a need is felt, I believe, for good political as well as psychological reasons: American democracy, if it is to survive within the super organizations of government and commerce, of industry and labor, is predicated on personal contacts within groups of optimal size -- optimal meaning the power to persuade each other in matters that influence the lives of each (20).

A community exists when shared values and interests provide an appropriate framework for a network of interdependent relationships. It is in such a context that a private life becomes meaningful, and without which, privacy is simply another form of isolation.

We need social institutions that reinforce and sustain the conditions under which community is possible. Our examination of the museum learning environment provides an excellent vantage point from which to appreciate the extent to which the quality of the public environment can determine the quality of private experience. These lessons become vital in their importance as human populations become increasingly dependent on public accommodation for the development of meaningful private experience.

REFERENCES

1. Ainsworth, M. D. S. and Bell, S. M. Attachment, exploration, and separation. *Child Development*, 1970, *41*:49-67.

2. Boocock, S. S. and Schild, E. O. (Eds.) *Simulation games in learning*. Beverly Hills, California: Sage Publication, 1968.

3. Bruner, J. S. *The process of education*. Cambridge, Massachusetts: Harvard University Press, 1961.

4. Bruner, J. S. *Toward a theory of instruction*. Cambridge, Mass.: Harvard University Press, 1966.

5. Carnegie Commission on Higher Education. *Toward a learning society; alternative channels to life, work, and service*. New York: McGraw-Hill, 1973.

6. Carr, S. and Lynch, K. Where learning happens. *Daedalus*, 1968, *97*:1277-1291.

7. Chase, R. A. Behavioral biology and environmental design. In M. Hammer, K. Salziner and S. Sutton (Eds.), *Psychopathology: Contributions from the biological, behavioral, and social sciences*. New York: Wiley-Interscience, 1972.

8. Chase, R. A. Information ecology and the design of learning environments. In G. Coates (Ed.) *Alternative learning environments: Emerging trends in environmental design and education*. Stroudsburg, Pa.: Dowden, Hutchinson and Ross, 1974.

9. Chase, R. A., Williams D. M. and Fisher, J. J., III. Design of play materials for infants. *DMG-DRS Journal: Design Research and Methods*, 1973, *7*: 294-305.

10. Chase, R. A., Williams, D. M. and Fisher, J. J., III. Exercises in the design of learning environments. In R. Ulrich, T. Stachnik and J. Mabry (Eds.) *Control of human behavior: In education* (Vol. III). Chicago: Scott, Foresman, 1974.

11. Chase, R. A., Williams, D. M., Welcher, D. W., Fisher, J. J., III, and Gfeller, S. F. Design of learning environments for infants. In S. T. Margulis, R. A. Chase and R. B. Bechtel (Eds.) *Privacy; social ecology; undermanning theory*. Milwaukee: Proceedings of the 5th Annual Conference of the Environmental Design Research Association, 1975.

12. Clarke-Stewart, K. A. Interactions between mothers and their young children: Characteristics and consequences. *Monographs of the Society for Research in Child Development,* 1973, *38*:6-7.

13. Coates, G. J. (Ed.) *Alternative learning environments.* Stroudsburg, Pa.: Dowden, Hutchinson and Ross, 1974.

14. Commoner, B. *Science and survival.* New York: The Viking Press, 1967.

15. Department of Education and Science. *Museums in education: Education survey 12.* London: Her Majesty's Stationery Office, 1971.

16. Dewey, J. *The public and its problems.* Chicago: Gateway Books, 1946.

17. Dewey, J. *Experience and education.* New York: Collier Books, 1963.

18. Dewey, J. *Democracy and education.* New York: The Free Press, 1966.

19. Dubos, R. *Man adapting.* New Haven, Conn.: Yale University Press, 1965.

20. Erikson, E. H. *Dimensions of a new identity: Jefferson Lectures, 1973.* New York: W. W. Norton & Co., Inc., 1974, pp. 123.

21. Fox, M. W. *Concepts in ethology; animal and human behavior.* Minneapolis, Minn.: University of Minnesota Press, 1974.

22. Haliday, M. S. Exploratory behavior. In L. Weiskrantz (Ed.) *Analysis of behavioral change.* New York: Harper & Row, 1968.

23. Harré, H. and Secord, P. F. *The explanation of social behavior.* Totowa, N.J.: Littlefield, Adams & Co., 1973.

24. Herron, R. E. and Sutton-Smith, B. *Child's play.* New York: John Wiley & Sons, 1971.

25. Huizinga, J. *Homo ludens; a study of the play-element in culture.* London: Routledge & Kegan Paul, Ltd., 1949.

26. Hutchins, R. M. *The higher learning in America.* New Haven, Conn.: Yale University Press, 1936.

27. Illich, I. *Deschooling society.* New York: Harper & Row, 1971.

28. Larrabee, E. (Ed.) *Museums and education.* Washington, D. C.: Smithsonian Institution Press, 1968.

29. Lloyd, K. E. Contingency management in university courses. *Educational Technology,* 1971, *11*:18-23.

30. Maddi, S. R. Exploratory behavior and variation-seeking in man. In D. W. Fiske & S. R. Maddi (Eds.) *Functions of varied experience.* Homewood Ill.: Dorsey Press, 1961.

31. Mason, W. A. Early deprivation in biological perspective. In V. H. Denenberg (Ed.) *Education of the infant and young child.* New York: Academic Press, 1970.

32. Millar, S. *The psychology of play.* Harmondsworth, Middlesex, England: Penguin Books, Ltd., 1971.

33. Reese, E. P. *The analysis of human operant behavior.* Dubuque, Iowa: Wm. C. Brown Co., 1966.

34. Schaefer, E. S. Need for early and continuing education. In V. H. Denenberg (Ed.) *Education of the infant and young child.* New York: Academic Press, 1970.

35. Schools Council. *Pterodactyls and old lace; museums in education.* London: Evans Brothers, Ltd., 1972. (Distributed in the U.S. by Citation Press, Scholastic Magazines, Inc., 50 W. 44th St., New York, N.Y. 10036).

36. Skinner, B. F. *Science and human behavior.* New York: The Free Press, 1953.

37. The American Association of Museums. *America's museums: The Belmont report.* Washington, D. C. The American Association of Museums, 1969.

38. Uzgiris, I. C. Ordinality in the development of schemas for relating to objects. In J. Hellmuth (Ed.) *Exceptional infant: The normal infant* (Vol. 1). New York: Brunner/Mazel, 1967.

39. Winstanley, B. R. *Children and museums.* Oxford: Basil Blackwell, 1967.

40. Wittlin, A. S. *Museums: In search of a usable future.* Cambridge, Mass.: MIT Press, 1970.

SOCIAL PRIVACY IN THE COMMUNITY OF DIVERSITY

Barrie B. Greenbie

Department of Landscape Architecture & Regional Planning
University of Massachusetts
Amherst, Massachusetts 01002

Synergy expresses the condition that the whole is greater than -- or at least qualitatively *other than* -- the sum of the parts. In a synergic view of communality and privacy, our collective selves, as a society, have an expanded identity which is not merely an aggregate of our personal identities, yet is composed of and dependent on them. Our relationship to the non-human physical environment, both as individuals and as groups, is also not merely additive. Larger spaces do not simply accommodate more people; social structures become qualitatively different entities as space expands or contracts.

For me, this relationship is best expressed through the concept of territory, defined as the physical expression of social form. Animal ethologists, who in recent years have brought a new perspective to this ancient subject, often make a distinction between *individual distance* and *territory*. For humans the former represents the privacy of the person. The latter is primarily a space for most important social activities, usually associated with a breeding unit among lower animals, and concerns the privacy of groups.

In addition, many ethologists make a distinction between territory, as a defended or exclusionary space, and *home range* which is an area simply used, but normally not defended, by a number of animals more or less known to each other. Paul Leyhausen has paid particular attention to this type of spatial organization, and he has observed in mammals very different kinds of social behavior and status ranking between defended home territory and neutral home

---

Acknowledgement: Part of this paper is condensed from B. B. Greenbie, *Design for Diversity*, 1976. Reproduced by permission of Elsevier Scientific Publishing Company.

range (11, 12, 13). He calls the two types of ranking systems *relative dominance hierarchy and absolute dominance hierarchy.* In the former the psychological advantage of one animal over another is related to the position of an encounter relative to the home base. A relatively low ranking weak animal can drive a much stronger more dominant one out of its own territory. In the neutral territory, status ranking and dominance power depend on individual characteristics, and once established in a group tend to remain for the life of the individuals involved.

To comprehend the meaning of territory for human beings, it is necessary to posit a much more complex model than a simple straight line bi-polar continuum with privacy at one end and communality at the other. In this paper I will try to summarize a model developed at considerable length elsewhere (4), which is four dimensional and hierarchical (Fig. 1), including the important dimension of time. It distinguishes between two kinds of privacy, the privacy of the individual vis à vis the primary group, and the privacy of the group vis à vis other groups in a rising order of complexity. It includes, and I hope usefully extends, Hall's concept of *proxemics*, which he defines as "the interrelated observations and theories of man's use of space as a specialized elaboration of culture." (9)

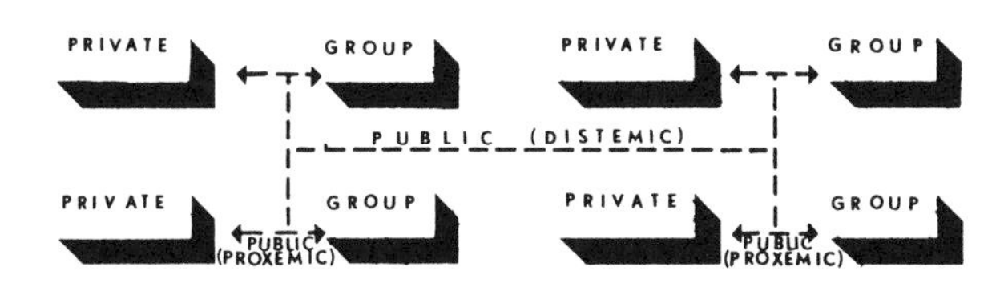

Fig. 1. Integrated democratic continuum

Hall considers proxemics to be the human extension of territorial and status ranking systems and the impulse to display, mate, fight, or take flight. The reason cultural differences in the proxemic use of space have not been widely recognized, according to his theory, is that they are largely out of awareness (9, 10). MacLean has found three layers in the brains of human beings which, in his theory, conform to various stages of evolution of the nervous system above the level of vertebrates (14, 15, Fig. 2). The most primitive of these, the *reptilian brain*, is similar in all vertebrates and governs environmental responses pertaining to basic species survival, including the more primitive expressions of territorial defense. We also share with other mammals a portion of the brain called the *limbic system* which, in this theory, governs emotional responses and

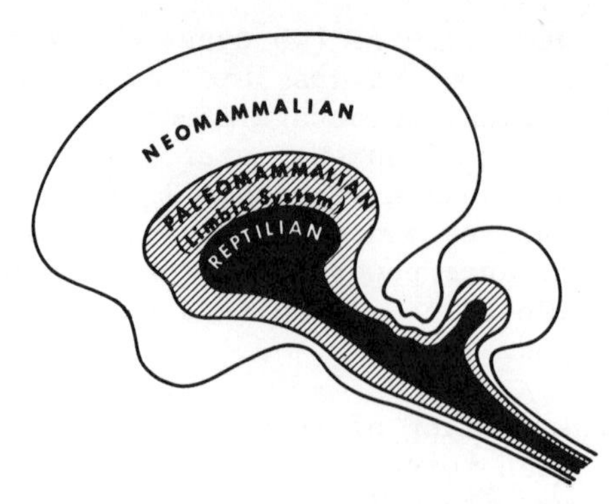

Fig. 2. In evolution the primate brain enlarges along the lines of three basic patterns that in hierarchic order may be characterized as reptilian, paleo-mammalian, and neomammalian, from MacLean (15).

enables mammals to develop much more complex forms of social organization. Surrounding this paleo-mammalian brain is the neo-cortex, which, most fully developed in man, contains the neural mechanisms associated with anticipation, planning, and verbal and mathematical language. These centers, while they operate as a unit which MacLean calls the *Triune Brain*, are partially independent and are different in chemistry, in structures, and in data processing capacities. In terms of MacLean's theory, we can conclude that what Hall defines as proxemic behavior relates mainly to the lower two of these three brain centers.

Esser has noted that these different brain centers, among other things, process information at very different speeds, with the new brain, through abstract verbal and mathematical concepts, able to appraise the environment much more rapidly than the older centers. He has also observed that spatial perception in the child develops before verbal language, and is therefore not wholly accessible to the new brain (3). We can, therefore, conclude that the spatial as well as the temporal paradigms of the newer and older brains are different.

MacLean observes that distinctions between emotional and rational behavior have long been made, but he feels that insufficient attention has been paid to the differences between behavior mediated by the paleo-mammalian emotional brain and the more primitive reptilian one (16). For the purposes of my model, however, the most significant distinction is between the phyletically new level which Esser calls the "intellectual brain," and which I am going to call the "planning brain," and the two lower ones which might, following Esser's reasoning, be called the "spatial brain." I asked MacLean if the two older centers could be treated as an entity in this way without unduly falsifying his theory, and he kindly suggested the term *draco-limbic* brain for this purpose. In the model I will present here, then, his "triune brain" will thus be converted to a "diune" one.

Essentially then we are concerned here with a cognitive distinction between our draco-limic human nature and our peculiarly human conceptual nature. This presents a paradox in considering the concept of "human scale" as applying to those relatively intimate, primary group spaces which most clearly fall within the province of proxemics. Literally, our most *human* nature is that which is at home in large reaches of space and time, which perceives events indirectly through the prostheses of technology. And yet, environments structured in such terms are often felt to be most unsatisfactory in the daily lives of most people; they are the very ones that proxemics would "humanize."

I believe we can find assistance with this problem by viewing proxemic culture as a complex and enriched extension of the animal's social territory, but also to allow for a qualitatively different socio-spatial context which is an elaboration of the animal's home range, made far more impersonal than the original by our technological and linguistic extensions through the new planning brain. Proxemics might thus be viewed as a *relative social hierarchy*, in Leyhausen's terms, *on cultural territories vis à vis other cultural territories*. Group members on their own turf have a sense of security which they do not have outside it. However, urban environments contain a great deal of space which is actively used but which cannot be considered anybody's turf. With our most recent technical extensions, the human range has expanded to include the entire planet and now is pushing out into the solar system. This is possible, in my model, precisely because the human conceptual brain is partly free of the draco-limbic brain, because it can operate somewhat independently as MacLean believes, and therefore is able to construct a cognitive world that is not limited by the more primitive personal types of social organization that bound our evolutionary ancestors to relatively small groups in more or less discrete spaces. This phylogenetically new cognitive world is, as far as the rest of our experience is concerned, very much like the artificial earth in which humans can now travel into space. Like the space capsule, we cannot reside in it for very long without returning to the original base on which we evolved.

The name I am giving this cognitive world is *distemic* to maintain continuity and establish a polarity with Hall's concept of *proxemic*. Both are a function of the use of space and therefore to talk of "proxemic *space*" is, strictly speaking, redundant, and so it is of "distemic space," but I am going to use the terms anyway to underscore their design aspects.

Hall has asserted that proxemics is concerned primarily with *context* as opposed to *information* (10). By *context*, he means the entire gestalt experience in a particular man-environment transaction, involving all appropriate senses in combination with relevant stored memories, largely out of awareness. *Information*, in his view, is based on abstract symbols with specific assigned meanings, consciously maintained. Information, in this sense, is a function of verbal (and presumably also mathematical) language, and is linear in structure rather than gestalt. Hall notes, of course, that no situation is entirely contextual or informational, and that people draw meaning out of the relationship between the two. But he feels that the relationship is generally inverse, that is, a high context situation is a low information one and vice versa.[1] The purpose of introducing the word distemics is to deal more adequately with the low-context high-information situations which he identifies but otherwise does not pursue very far.

The basic practical difference between proxemic space and distemic space, from a planning and design perspective, is that whereas proxemic use of space is culturally determined, distemic space can be primarily trans-cultural, or super-cultural. One can fully learn to be a German, an Arab, or an American only by being born one. Much of the learning takes place in the infant dawn of consciousness. With sufficient good will and discipline, proxemic groups can learn to tolerate, to respect, to accommodate, and in some cases actively enjoy each others' diversities, but only in the most limited ways to experience them. By contrast, in a relatively short time a fully adult member of any culture with the right mental and physical equipment can learn to be a physicist at home in a laboratory, a pilot hopefully in unambiguous communication with a control tower, or a stevedore doing a first rate job with a multi-ethnic crew on the docks. In all three cases very specific behavioral systems involving the use of space, with clear understandings as to the conventions of its use, are taking place in spaces designed for such behavior. Distemic behavior in distemic space ideally accommodates proxemic differences without infringing on them by making them *essentially irrelevant in that context.*

Distemic space and proxemic space are by no means mutually exclusive. The same space may perform both functions, or one of the functions for one group and not the other. For instance, a traditional street market in an ethnic neighborhood of a cosmopolitan city, or a village frequented by tourists, may be largely proxemic for its proprietors and largely distemic for visitors. On the other hand, it must be partly distemic even for its proprietors who will have to deal with alien behavior to an extent far greater than they would be likely to tolerate comfortably in their homes. But it may

[1] See also, Hall, E.T. *Beyond Culture*. Garden City, N.Y.: Doubleday, 1976.

also have proxemic value for the visitors who vicariously enjoy the foreign culture as culture, in contrast to the more objective activity of selecting a commodity. There is no presumption in this postulate that human beings always, or even usually, perceive either activities or space exclusively in proxemic *or* distemic terms. As noted, both types of environment-behavior relationships may exist simultaneously in the same space or place. Since the proxemic-distemic relationship is a continuum, various groups or individuals may occupy various points on it and different times, and one set of relationships can be transformed into the other. Distemic relationships which exist for a long enough period of time (more than a generation) may (and possibly must) become proxemic in their own right. This is probably one way in which cultural evolution occurs.

Perhaps both the distinctions and the possibilities for convergence can be illustrated by assuming four different kinds of recreational settings prevalent in the United States: (1) a pool hall in an ethnic working class neighborhood, (2) a bowling alley in a shopping center, (3) a bridge club, and (4) a golf course in an exclusive country club. The first will be organized almost entirely along ethnic lines, or a narrow mixture of proxemically similar ethnic cultures, such as perhaps second or third-generation-American-Polish and Italian-Catholics. The second may include a broader ethnic spectrum, but be proxemically limited by class, say lower middle. The third may be ethnically quite mixed, but be proxemically limited by class, say suburban upper middle class. The last may be limited in terms both of class and ethnic origin, say upper class white Anglo-Saxon Protestants. Each of the games played will have two components, one the proxemic behavioral systems identified by Hall, the other the objective rules of the game. Any of the groups can readily learn the rules of any of these games and play them. However, the proxemic behaviours that make up the entire gestalt in each of these settings are not readily transferrable. A member of one who attempts to join the others on this level may at least to some extent inhibit or embarrass the group or be inhibited and embarrassed in it, and may invoke covert or even overt resentment.

Now let us shift the scene to Yankee Stadium in New York City. All of these groups may very happily enjoy the ball game, booing or cheering in harmony, and generally enjoying each other's company without much inhibition. Behavior which would be a rude breach of etiquette in any one of the other four settings, will be cheerfully tolerated or ignored. A common understanding of the rules of the game being watched (a conceptual process) combines with the primitive infra-cultural excitement of competition and aggressive, cooperative teamwork to create a heterogeneous functional social unit. This is what I mean by a distemic situation and distemic space. However, the passionate attachment of many Americans to baseball, with its special jargon, and certainly some of the behavior patterns of ball players, have through time also begun to acquire proxemic meanings which differ, for instance, from the equivalent European proxemic identification with soccer. Even so, such large scale spectator sports will tend to remain toward the distemic end of the continuum, because with relatively little instruction in the logical rules of the games, people of any nation and social class can quite easily share the enthusiasm.

While distemic situations are by definition those not *primarily* determined by life-long cultural considerations, they nevertheless do not exclude the emotions and a wide range of gestalt feeling states. Just as purely conceptual functions are not ruled out of proxemics, neither are draco-limbic ones ruled out of distemics. As creative scientists well know, there is an emotional color and excitement in the pursuit of abstract ideas that can rival in intensity and pleasure, (as well as fear and depression) more "paleo-mammalian" activities. In fact, distemic space unifies human beings on two levels: on the level of pre-cultural experience as well as super-cultural experience. All artists (including poets, novelists, and dramatists who use words for evocative instead of descriptive purposes) work out of the centers of gestalt, non-verbal experience on a level which in some respects transcends culture, presumably by touching those substrates of experience which are innate and species-specific. The best art of any culture will find response in the more open members of any other, even though creators and responders could not live together on close terms. Thus the Louvre and the Metropolitan Museum of Art, the Vienna National Opera and the New York Carnegie Hall are *primarily* distemic rather than proxemic. This is extremely important as we move from planning to design.

I should note that Hall has provided for situations where different proxemic patterns interact, using a complimentary term of his own, *proxetics* (10). However, as he presents it, proxetics appears to define an approach to observing or analyzing proxemic behavior, and not the overarching behavioral situations which I am calling distemic. I was not aware of Hall's new word when I developed my own concept, and I shall stick to the formulation for the present. Among other things, the prefixes have a much stronger *spatial* connotation, relating to *proximate* and *distal*, a contrast in scale, which, as I perceive it, applies not only to geographical but also to psychological and social distance. The differences between proxemics and distemics in my model are similar to the primary group and secondary group relationships traditionally identified by sociologists. But they are not identical, for a number of reasons which space does not permit me to elaborate here, except to note that proxemics can include both, whereas distemics will usually involve only secondary ones.

I believe these distinctions are not merely academic or esthetic, but life and death matters. Cassel has shown that physical disease correlates very strongly with the breakdown of social organization which often accompanies migration, especially under circumstances of enforced relocation (1, 2). Cassel finds it also accompanies loss of status. It often is a concomitant of crowding, but not always so, and his theory is that social disruption, not crowding *per se*, causes the problems often blamed on crowding alone. Thus the preservation of proxemic territories becomes essential for physical health, and this is likely to be even more critical for those lowest in status and most economically deprived who do not have resources to protect group territories on their own. Any satisfactory scheme for organizing real world space must take into account not only distemic space and proxemic space, but also provide *relative security and freedom* within those spaces. Proxemic space, if my hypotheses are valid, will be concerned with *security-freedom relationships within the* small group, whereas distemic space will consist primarily of those relationships *between groups* through which individuals move as individuals, structured in terms of two continua, not one, corresponding to the two kinds of cognitive space.

In conclusion, I will apply this model to a community which was recently the focus of a small pilot study (6, 8). The purpose of the study was to explore various methodological problems in testing an hypothesis, which I had generated from Cassel's work, that physical and social health would correlate with the presence of community boundaries which define and protect stable social structures (7). The study was quite inconclusive regarding its main objectives, but it produced some useful information which supports and illustrates the model presented here.

In the attempt to develop a measure of community boundedness we mapped the perceptions of five different sets of subjects, including ourselves, the researchers, eight taxi drivers, eleven public health nurses and 89 residents of three areas. We also took the official "planning district map" as a measure of planner's perception. The great variety of perceived boundaries in these areas is shown in Fig. 3 - 7. (For more details see 4).

Of the three areas studied, the area due east of the Central Business District, variously called "Winchester Square" or "Hill McKnight," had few consistently defined boundaries, although it was full of barriers which were also the least coherent set of physical edges in of all the study areas. The area known variously as "The North End" or "Brightwood" was somewhat more consistently defined by all respondents, as it also was more coherently bounded physically, although it was badly lacerated by a railroad, interstate highways, and urban renewal, and was in the process of radical population change. However, the area called the "South End" was nearly unanimously identified as a neighborhood by all five subject groups,

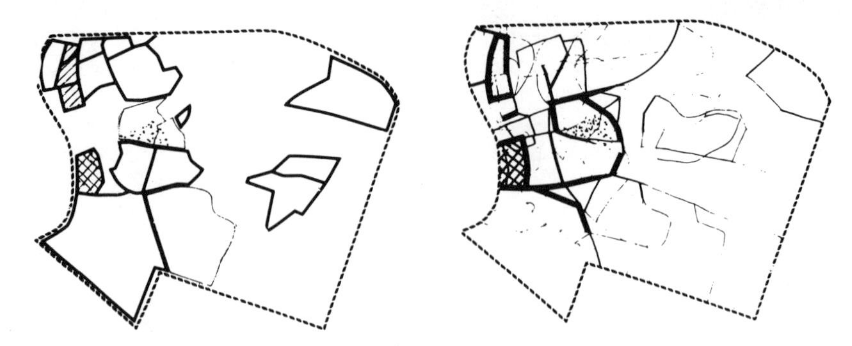

Fig. 3. Neighborhood boundaries in Springfield as perceived by researchers.

Fig. 4. Neighborhood boundaries in Springfield as perceived by taxi drivers.

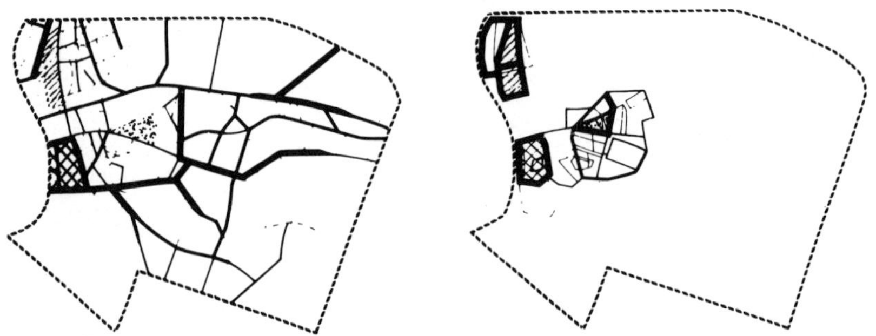

Fig. 5. Neighborhood boundaries in Springfield as perceived by public health nurses.

Fig. 6. Neighborhood boundaries in three areas of Springfield as perceived by residents.

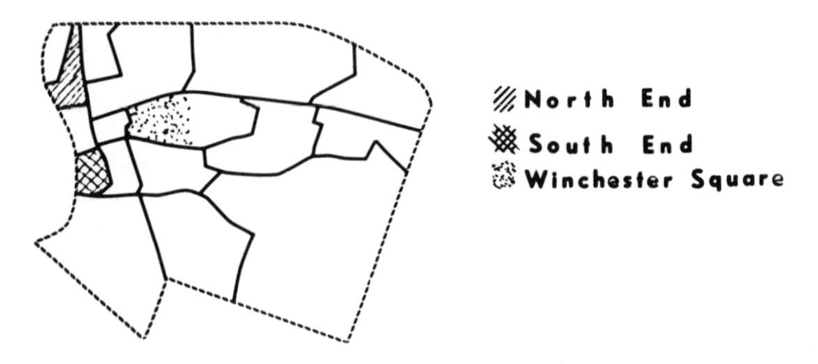

Fig. 7. The officially recognized "planning districts" in Springfield. (Figs. 3-7 redrawn by Margaret Kent from authors' originals.)

including the Planning District Map, and by that single name. This was also an area of almost total agreement on physical boundaries. The predominantly residential parts of it consist of a series of dead-end streets running off a major commercial artery, connecting directly on the north with the central business district. It is bounded on the west by steep bluffs, and on the south by a hill, a T intersection, and a statue of Columbus to which all respondents referred (see accompanying photographs). The "South End" is a traditionally Italian ethnic area. Some residents mentioned a small sub-area with particular enthusiasm as "a good place to live." It is an architecturally undistinctive but very cloistered area a few blocks square, called "Hollywood," which, as the photographs suggest, is visually enclosed in all directions with a small commercial center of its own. It very closely fits Newman's description of a "defensible space" (17), and its defensibility was brought home to us by the fact that while we have been conspicuously photographing all the other areas without drawing much interest, in this area our cameras attracted inhospitable attention, as did we ourselves, and one proprietor angrily refused to let himself or his shop be photographed. Subsequent information confirms that the area is considered a good safe place to live, and that the residents are quite concerned that it remain so.

One conclusion is that the availability of clearly bounded physical areas will increase the likelihood that homogeneous populations already predisposed toward social cohesion, such as ethnic groups, will seek them out in proportion to the degree to which ethnic or other proxemic community identity overrides other, possibly distemic, needs and objectives. I think this can be supported by a look at the environments of most traditional, stable ethnic communities in any of our cities. For instance, Boston's Italian North End and Irish South Boston are both essentially peninsulas, now separated from the rest of the city in the first instance by an elevated highway and in the second by a bridge. The same is true of Bayonnne, New Jersey.

But the other side of the proxemic-distemic relationship is also suggested by our study. The highly bounded, socially homogeneous, ethnic South End did not lack distemic spaces and relationships (Fig. 8). It is in no sense a ghetto. On the contrary, it revealed these much more than the other two areas, both internally and externally. The South End has its own distemic space in the Main Street spine, a variety of commercial establishments and some industry. The proxemic aspects of this street are observable in its overall ethnic flavor, an Italian market, restaurants, bars, and lunch counters. This might also be called macro-proxemic. The street is a minor arterial connecting directly with the CBD, which has been renovated through unusually successful and lively urban renewal projects. In fact, the distemic heart of the city is the northern boundary of

The South End of Springfield

"Hollywood" in the South End

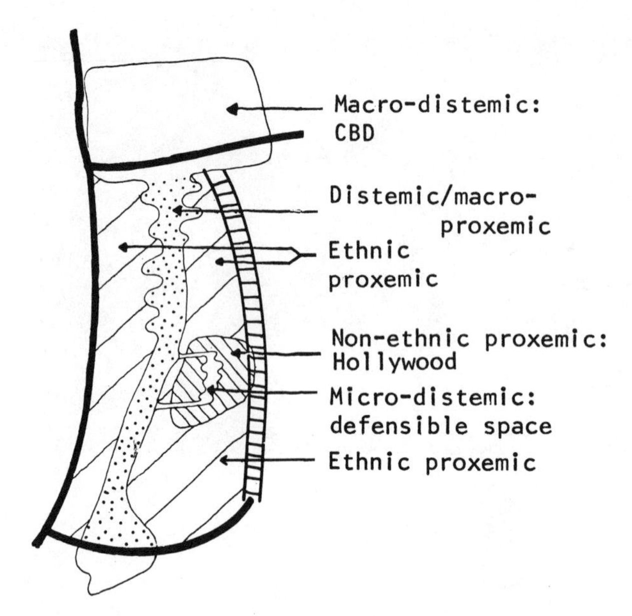

Fig. 8. The proxemic-distemic hierarchy in Springfield's South End. (From: B. Greenbie, *Design for Diversity*. Elsevier Scientific Publishing Co., reproduced by permission.)

this community, and the CBD can be called macro-distemic in this hierarchy. At the other end of the scale, the sub-community of Hollywood has its own main street, with some minor service type shops and a more or less mini-scale distemic center, the Hollywood Cafe. The proxemic character of this area has little visible ethnic quality, not nearly as much as some of the other residential cul de sacs, but rather it appears to be a community of limited liability as in Suttles' model (19). Taking the South End as a whole, there is an hierarchical relationship of proxemic bounded areas with distemic spaces increasing levels of trans-cultural interaction.

The North End apparently had at one time a somewhat similar set of relationships. Its Memorial Square (see photos) had constituted a social and cultural, rather than commercial, distemic center for the community, with a church and library and a nodal arterial intersection, with North Main Street linking it too with the CBD. That link has now been interrupted by an east-west expressway that constitutes a visual barrier on the one hand, and deflects traffic away from the area on the other. Memorial Square is no longer a center, its cultural symbols, to the extent that they survive at all, belong to a community population now isolated from it to the north. The church maintains an active congregation, but most of the parishioners now live elsewhere. The new and/or transient population in

Memorial Square in Springfield's North End.

the immediate area have so little to do with this group that the church plays no distemic role for them. Only a few respondents mentioned the church as a landmark, and, whereas in the South End all respondents mentioned the Statue of Columbus, here no one mentioned an equally prominent statue. The abandoned library was mentioned with bitterness by some older residents.

In the Winchester Square area, distemic relationships are fragmented and proxemic ones ghettoised by strong barrier that are neither socially or physically functional. The square itself has remained a somewhat marginal commercial industrial node, but for a number of reasons, including heavy through traffic on State Street, it is more of a barrier than a distemic center. Since our study was conducted, a change in traffic routing has aggravated this problem, forcing some of the remaining establishments out of business, even though the area had been a target for Model Cities and other rehabilitation efforts which in some sections of the area have been more or less successful.

I and my students at the University of Massachusetts are now attempting to apply this model to other communities (4), both existing and planned. It will probably be easier to identify these relationships in older existing communities than to design them *a priori* into new developments. As a planner, I view this theory, like most other theories, as a framework for asking questions and obtaining information, rather than as a predictive tool. In every case, satisfactory community design calls for attention to particularities. Here, I think, is where the behavioral scientist, the generalized planner, and the designer can best develop their synergetic associations. By accepting different roles directed to different classes of problems based on the different perceptual systems of our dichotomous brain we can perhaps minimize the conflicts that bedevil so many interdisciplinary associations (18). To the designer, however, falls the most creative task of linking the planning and the spatial brain, while providing a complex hierarchical habitat that can accommodate both cognitive worlds.

REFERENCES

1. Cassel, John. Health consequences of population density and crowding. *Rapid Population Growth*, National Academy of Sciences, Baltimore Md., Johns Hopkins Press. Ch. 12. Reprinted in *People and Buildings*, Robert Gutman (Ed). New York: Basic Books, 1972.

2. Cassel, John. Physical illness in response to stress. In Sol Levine and N.A. Scotch (Eds.) *Social Stress*. Chicago: Aldine-Atherton Press, 1970.

3. Esser, A. H. (Ed.) *Behavior and Environment.* New York: Plenum Press, 1971.

4. Greenbie, Barrie B. *Design for Diversity: Planning for Natural Man in the Neo-Technic Environment.* New York: Elsevier Scientific Publishing Company, 1976.

5. Greenbie, Barrie B. *Jamestown, N.Y., A Social/Spatial Profile.* Amherst, Mass.: The University of Massachusetts, 1976.

6. Greenbie, Barrie B. Contrasting and consistent perceptions of neighborhood boundaries in a New England City. *Man-Environment Systems,* 1975, *5*(5):327.

7. Greenbie, Barrie B. Social territory, community health, and urban planning. *Journal of the American Institute of Planners,* 1974, *40*:74-82.

8. Greenbie, Barrie B., Tuthill, Robert W. and Brown, Marilyn A. Contrasting cognitive maps of city neighborhoods by diverse segments of the population. Unpublished manuscript, 1973.

9. Hall, Edward T. *The Hidden Dimension.* New York: Doubleday, 1966.

10. Hall, Edward T. *The Handbook of Proxemic Research.* Washington, D.C.: Society for the Anthropology of Visual Communication, 1974.

11. Leyhausen, Paul. The communal organization of solitary mammals. *Symposium,* 1965, *14*:249-63. London: Zoological Society. Reprinted in Harold M. Proshansky, et al. (Eds.) *Environmental Psychology.* New York: Holt, Rinehart & Winston, 1970.

12. Leyhausen, Paul. Dominance and territoriality as complemented in mammalian social structure. In A. H. Esser (Ed.) *Behavior and Environment* . New York: Plenum Press, 1971.

13. Lorenz, Konrad and Leyhausen, Paul. *Motivation of Human and Animal Behavior.* New York: Van Nostrand Reinhold, 1973.

14. MacLean, Paul D. The Brain's Generation Gap: some human implications. *Zygon/Journal of Religion and Science,* 1973, *8*:113-127.

15. MacLean, Paul D. The brain in relation to empathy and medical education. *The Journal of Nervous and Mental Disease,* 1967, *144*:374-382.

16. MacLean, Paul D. On the evolution of three mentalities. *Man-Environment Systems*, 1975, *5*:213-224.

17. Newman, Oscar. *Defensible Space*. New York: Macmillan, 1972.

18. Sanoff, Henry. On EDRA purpose. *Design Research News*, 1974, *1*:3.

19. Suttles, Gerald D. *The Social Construction of Communities*. Chicago: The University of Chicago Press. 1972.

Contributors

John R. Aiello is Assistant Professor of Psychology at Douglass College of Rutgers University. He did his undergraduate work at City College of New York, received a Masters degree from Queens College of C.U.N.Y. and obtained his Ph.D. from Michigan State University. He is a member of the editorial board of *Environmental Psychology and Nonverbal Behavior* and has served as Associate Editor of *Human Ecology*. His current research is concerned with the effects of crowding and proxemic behavior.

Charles Baran went to the University of Toronto for his undergraduate training in psychology. After receiving his Bachelor's degree in 1975, he went to Memorial University in Newfoundland, where he was awarded a Masters degree in 1977.

Robert Beck is an environmental psychologist with a Ph.D. from the University of Chicago. Dr. Beck carried out research on pedestrian density for the Regional Plan Association in New York, studied tourist images and mental maps of European cities while teaching in the School of Geography at Clark University. Beck taught for four years in the Ecole d'Architecture, Université de Montreal, where he participated with colleagues in the design research of housing, mental health centers and office buildings. Currently, he is working as a television producer and researcher in the area of community participation in political programming and is a research consultant on the Urban Homesteading of Abandoned Housing to a Deputy-Mayor of New York City.

Alan Booth is Chairman of the Department of Sociology, the University of Nebraska-Lincoln. He received his B.A. from Antioch College and his Ph.D. in sociology at the University of Nebraska-Lincoln. For two years Dr. Booth conducted a study of urban crowding in Toronto for the Canadian government. His interests are in the area of the effect of the built environment on social relations and the quality of life.

Richard Allen Chase received his A.B. degree from the College of the University of Chicago and his M.D. degree from the Columbia University College of Physicians and Surgeons. He is presently Associate Professor of Psychiatry and Behavioral Sciences at The Johns Hopkins University School of Medicine and Professor of Architecture in the College of Architecture and Environmental Design, Virginia Polytechnic Institute and State University. Dr. Chase is interested in behavior ecology and environmental design, and he has done original research on the ways in which environmental design can optimize opportunities for learning and social interaction.

Devra Lee Davis obtained her B.S. and M.A. at the University of Pittsburgh and her interdisciplinary Ph.D. at the University of Chicago. After having taught in the Sociology Department, Queens College, New York City, she became Executive Director of the Health Promotion Project of World Man Fund, Washington, D. C. Subsequently, she served as Special Assistant for Toxic Substances and is presently Executive Secretary of the Administrative Toxic Substances Advisory Committee of the Environmental Protection Agency. Dr. Davis is interested in the area of health-related behavior, including the use of acupuncture and the development of science policy in the regulatory framework.

Alton J. DeLong has degrees in psychology, linguistics and man-environment relations (a Ph.D. from the Pennsylvania State University) and has a background in architecture and anthropology. He is presently Associate Professor in the School of Architecture at the University of Tennessee, where he teaches proxemics, design and man-environment relations. His research interests include space-time relationships in designed environments, scale-model simulation of behavior and the man-environment relationship as a process of communication.

Michael Efran obtained his Ph.D. from the University of Texas at Austin and taught psychology at the University of Toronto from 1969 to 1976. Dr. Efran is author of a book, *General Psychology*, as well as numerous journal articles dealing with man-environment issues and other topics in personality and social psychology. He is currently a consultant to government and industry and lives in Toronto.

Yakov M. Epstein is Associate Professor of Psychology at Rutgers University. He did his undergraduate work at the University of Pennsylvania and obtained his Ph.D. in social psychology from Columbia University. He is a member of the editorial boards of *Environmental Psychology and Nonverbal Behavior*, *Population: Behavioral, Social and Environmental Issues* and is presently Associate Editor of the *Journal of Applied Social Psychology*. He has served as Eastern Regional Coordinator of the APA task force on environment and behavior. Together with Andrew Baum, he is co-editing *Human Response to Crowding*. His current research focuses on the effects of crowding.

Aristide H. Esser is Director of Psychiatric Services at the Mission of the Immaculate Virgin, a child care agency in Staten Island, New York. After obtaining his M.D. at the University of Amsterdam in 1955, he specialized in psychiatry at the University of Leyden, The Netherlands. After ten years of research at Yale University, the Rockland Research Institute and the Mental Retardation Research Unit in Letchworth Village, both in the State of New York, he developed Central Bergen Community Mental Health Center in Paramus, New Jersey, which terminated in 1977. He was instrumental in founding the Association for the Study of Man-Environment Relations and is Editor of its journal, *Man-Environment Systems*, as well as its *M-ES Focus* monograph series. Dr. Esser has published articles and films and edited *Behavior and Environment* (1971) for Plenum Press.

Gary W. Evans is Assistant Professor of Social Ecology at the University of California, Irvine. He has a Ph.D. in psychology with emphasis in perception and environmental cognition. His primary research interests are spatial behavior and environmental cognition. He teaches courses in these areas as well as environmental psychology and research methodology.

Barrie B. Greenbie received his M.S. and Ph.D. in Urban and Regional Planning from the University of Wisconsin. He taught at Skidmore College and the University of Wisconsin and is currently Professor of Landscape Architecture and Regional Planning at the University of Massachusetts. His career spans thirty years of multidisciplinary professional and academic activity covering art, drama, architecture, applied engineering, industrial design, town planning, research and teaching. Throughout, the relation of behavior to space has been his guiding interest, along with the integration of technic and natural form. Dr. Greenbie authored journal articles and planning reports on environmental and behavioral subjects; his latest book is *Design for Diversity* (New York: Elsevier, 1976). His architectural designs have been published in the U.S.A. and Europe. His applied research includes several patents on building systems related to preservation of social and natural environments.

Hein Jaanus is on the faculty of the Department of Psychology at the University of Amsterdam. He obtained his Ph.D. in Social Science at this university in 1954. His research interests are environmental and social psychology, centered around learning and motivation, e.g., as applied to behavior in man-made environments. For this purpose, he joined Bob Nieuwenhuijse in research on the analysis of perception and evaluation of residential areas.

Robert A. Karlin did his undergraduate work at Harvard University, obtained a Masters degree in Psychology from Yale University and a Ph.D. in Clinical Psychology from Rutgers University. His doctoral thesis dealing with the effects of acute experimental crowding was awarded honorable mention in a contest sponsored by the Society for Experimental Social Psychology. After completing his degree, he served as Director of the Alcohol Behavior Research Laboratory at Rutgers University. Following this, he joined the faculty of University College, Rutgers University, where he is presently Assistant Professor. Dr. Karlin is a licensed clinical psychologist specializing in behavior modification and the use of hypnosis for pain relief. His recent work concerns the effects of crowding and the nature of hypnosis.

Bob Nieuwenhuijse is an Associate Professor in the Department of Psychology at the University of Amsterdam. He obtained his Ph.D. in Social Sciences at this university in 1969. His research interests are in general and environmental psychology.

Walter Ohlig is a graduate student in the Program in Social Ecology at the University of California, Irvine. He was born and received his early education in Germany. While pursuing a career in Data Processing, he designed a computer system for a public housing program, a HUD-funded traffic safety system and municipal land-use planning applications. His interest in urban planning brought him to UCI's Program in Social Ecology, where he earned a Bachelor of Arts degree. His major research interests are in the areas of environmental analysis and design, with special emphasis on environmental psychology.

Susan M. Resnick is Coordinator of Research and Evaluation for Orange County Revenue Sharing Social Programs. Her current research focuses on the application of social science knowledge and methods to the formation of public policy. Ms. Resnick received a B.A. in Psychology from U.C.L.A. and an M.A. in Social Ecology from the University of California, Irvine, where she has taught research design.

Daniel Stokols is on the faculty of the Program in Social Ecology at the University of California, Irvine. He received his doctorate in social psychology at the University of North Carolina, Chapel Hill. His current research interests include human crowding, the behavioral and health consequences of exposure to traffic congestion and the utilization of behavioral research in the design process. Dr. Stokols recently edited *Perspectives on Environment and Behavior: Theory, Research and Applications* (1977) published by Plenum Press.

Pierre Teasdale is an architect trained at McGill University and is Associate Professor of Architecture in the Ecole d'Architecture, Universite de Montreal. Teasdale is responsible for the Man-Environment Relations program at the university. He has carried out extensive design research on a variety of building forms, including schools and housing, and on such transportation forms as airport control towers. Currently, Teasdale, in association with Robert Beck, is preparing design guidelines for residential roof decks for the Central Mortgage and Housing Corporation, federal government of Canada.

Allen Turnbull recently completed his doctoral work at Carleton University. The research reported here is based on a portion of his Ph.D. thesis on privacy, community and activity space. The author's research interests focus on the relationship between the built environment and mental health.

Craig M. Zimring is a doctoral student in environmental psychology at the University of Massachusetts at Amherst. He is currently serving as Co-Director of the ELEMR Project, a longitudinal study of environmental behavior at a state school for the developmentally disabled. His specialties include environmental cognition, multidisciplinary research methods and evaluation of the built environment.

Ervin H. Zube is Director of the Institute for Man and Environment at the University of Massachusetts in Amherst. He received a Master of Landscape Architecture degree from Harvard University and a Ph.D. in Geography from Clark University. He has served on the landscape architecture faculties at the University of Wisconsin, University of California, Berkeley and at the University of Massachusetts at Amherst, where he served as department head. His current research activities are related to man-environment studies and focus on evaluation of the designed environment and on environmental perception.

AUTHOR INDEX

Subject Index